John Browne

A compleat treatise of the muscles, as they appear in humane body, and arise in dissection:

With diverse anatomical observations not yet discover'd : illustrated by near fourty copper-plates, accurately delineated and engraven

John Browne

A compleat treatise of the muscles, as they appear in humane body, and arise in dissection:
With diverse anatomical observations not yet discover'd : illustrated by near fourty copper-plates, accurately delineated and engraven

ISBN/EAN: 9783337713683

Printed in Europe, USA, Canada, Australia, Japan

Cover: Foto ©ninafisch / pixelio.de

More available books at **www.hansebooks.com**

A Compleat Treatise
OF THE
MUSCLES,
As they appear in
Humane Body,
And arise in
DISSECTION;
With Diverse
Anatomical Observations
Not yet Discover'd.

Illustrated by near Fourty
COPPER-PLATES,
Accurately Delineated and Engraven.

By John Browne, *Sworn Chirurgeon in Ordinary to His Majesty*.

Non Nobis Nati.

In the *SAVOY*.
Printed by *Tho. Newcombe* for the Author. 1681.

TO HIS
Moſt Sacred Majeſty
CHARLES II.
By the Grace and Providence of God
KING
OF
Great Brittain, France, and Ireland,
Defender of the Faith, &c.

THESE

Anatomical Exercitations

OF

MUSCULAR DISSECTION,
Are moſt Humbly Dedicated, by
Your Majeſties
Moſt Loyal Subject,
And moſt Obedient Servant,
John Browne.

Charles R.

CHARLES *By the Grace of God, King of* England, Scotland, France, *and* Ireland, *Defender of the Faith,* &c. *To all Our loving Subjects of what Degree, Condition, or Quality soever within Our Kingdom and Dominions, Greeting: Whereas it hath been manifested unto Us, that Our Trusty and Well-beloved* John Browne *Esq; one of Our Chirurgeons in Ordinary, hath not only with great Art, but at the Expence of much Time and Charge, delineated, described, and accurately Engraved in Copper-plates an* Anatomical Treatise of Muscular Dissection; *which performance of his is to Our great liking and satisfaction, so that We may express Our Approbation thereof, and give him all due and ample encouragement for the future; We do hereby signifie Our Royal Pleasure, granting unto the said* John Browne *the sole Priviledge of Printing the aforesaid Treatise with its Copper Figures; and strictly Charging, prohibiting and forbidding all Our Subjects to Copy or Counterfeit any the Sculptures or Description aforesaid, either in great or small, or to Import, Buy, Vend, Utter or Distribute any Copies or Exemplars of the same Reprinted beyond the Seas within the term of Fifteen Years next ensuing the Date of this Our Licence, and Prohibition, without the Consent and Approbation of the said* John Browne, *his Heirs, Executors and Assigns, as they and every of them so offending will answer the*

the contrary at their utmost *Perils*; *Whereof as well the Wardens and Company of* Stationers *of Our City of* London, *the Farmers, Commissioners, and Officers of Our Customs, as all other Our Officers and Ministers whom it may concern, are to take particular notice that due Obedience be given to this Our Royal Command.*

Given at Our Court at *Whitehall* this 22th Day of *Nov.* in the Three and thirtieth Year of, *&c.*

By His Majesties Command,

L. Jenkins.

TO

To His Grace
CHRISTOPHER
Duke of Albemarle,

Earl of Torington, Baron Monck of Potheridge, Beauchamp, and Teyes, Knight of the Most Noble Order of the Garter, One of the Gentlemen of His Majesties Bed-Chamber, One of His Majesties most Honourable Privy-Council, Lord Lieutenant of the Counties of Devon and Essex, And Captain of His Majesties Guards of Horse.

May it please Your Grace,

IS *Majesty having been Graciously pleased not only to take a view of, but to allow the Publication of this Treatise of* Muscular Dissection ; *Now after Him it waits at Your Graces Feet, imploring Your Pardon for its Presumption : But Your Generous Spirit always encouraging the Industrious Man, breathes Life into my Undertaking, by which Your Grace does declare to the World Your valuable Goodness, as well as Your substantial Honour : You never exercising Your self in little Designs, but in the Publick Service of Your King and Countrey ;*

The Epistle Dedicatory.

Countrey; These being the main Columnes of Your Great Lustre, all which when Greatness shall be shrivel'd into nothing, or at least into a cold Remembrance, will raise Yours into everlasting Praise, and force future Ages to speak well of Your Merit.

And were my Performances equal to my Wishes, I might not doubt of a happy Reception; but since the knowledge of my own weakness forbids the entertainment of such thoughts, it waits on Your Grace with all Humility: And whilst You have so much of that Heaven about You, I shall fear little dammage from any Earthly defects of my own.

*And Noble Sir, As You are not only the true Heir of the Fortunes, but of the incomparable Worth of so Great a Father, and so Renowned a General, whose Character being too big for my Pen, I dare leave it to any Impartial Reader of this Age to Judge, whether You do not most worthily participate with him, both in Your Benign Temper and Honourable Conduct; The delineating of both whose Heroick Spirits, may well challenge the best of Orators to describe; Whose Goodness joined with most noble Qualifications, may as deservedly Entitle You both to that which was once given to the great Emperor Titus: (*THE DELIGHTS OF MANKIND.*) But why should I strive to tell the World what they already know, and what I am confident none will Dispute? 'Tis an Honour too great for me, that Your Grace hath (by laying this Command upon me) given me an occasion to publish the hearty desires I have to acknowledge all Your Graces Favours, and to assume a liberty (few Men living but will be proud to share in) of declaring my self,*

Your Graces,

Most Humble, most Faithful,

And most Obedient Servant,

John Browne.

Librum hunc (cui Titulus Myotomia) Dignum Judicamus, Qui Imprimatur,

Al. Frasier M. R.
Charl. Scarburough M. R.

Johannes Wicklethwait Præses Colleg. Medicor. Londinens.
Thomas Witberley Censor.
Samuel Collins Censor.
Tho. Millington Censor.
Edvardus Browne Censor.

Clarisſimis, Doctiſſimiſque Viris D. D. Præſidi & Sociis Collegii Regalis Medicorum Londinenſium, Myologicam hanc exercitationem ex animo lubens ac merito Dicat Dedicatque, J. Brown.

Quantum Lucis Scientiæ conferat Methodica Inſtitutionis ratio, non vobis ſolummodo (Literatorum Coryphæi) ſed & Tyronibus etiam omnibus tàm clarum eſt, quam quod Clariſſimum. Partes enim ſi Debitâ ſerie tractentur explicenturque, & perſpicué abinvicem dignoſcuntur, & multó melius a Lectoribus percipiuntur; Quippe quod Ordinata rerum diſpoſitio præcepta, tàm Jucundiora, tàm faciliora reddat, dum ſingula ſeparatim in ſuâ Claſſe repoſita, memoriam adjuvant & oblectant. Ut igitur iſta tam manifeſta prætermittam, de re ipſâ dicere liceat. Antiquiores hanc Scientiam Anatomicam in duas partes diviſerunt, Actionem, & Contemplationem, quarum alterius ope, quicquid in Cælo, Terrâ Marique videri uſquam poſſit, in uno homine conſpicitur, & tot miracula deteguntur, ut Calamus iis enumerandis oneri ſuccumberet, & admiratio nos penitus obruat. Nihil eſt quod vel à Naturâ, vel Arte expectari poſſit, quod hæc Ars non proferat. Subjectum enim ejus adeo nobile eſt, nimirum Corpus hominis, ad Imaginem Dei, à Deo creati, ut nihil nobilius ſub Sole inveniri poſſit; & ſingulas humani corporis particulas adeo eleganter deſcribit, ut Arti ſimul veſtræ Medicorum, & noſtræ Chirurgorum

rurgorum summo ornamento fiet. Veritatem in abditis musculorum (& cæterarum partium) meatibus & tenuissimis fibrillis latentem è Tenebris vindicat, & in clarissimam Lucem revocat. Obscuris Lumen, Obsoletis nitorem, Fastiditis gratiam, Dubiis fidem, & vetustate quasi abrogatis Authoritatem conciliat: Per hanc Medici ferendo Judicio aptiores fiunt; cum Symptomatibus & indicationibus instructi citius certiusque affirmare queant, quam partem, quâ causâ morbi dolor, urgeat, afficiatque imò ex ipsâ mortuorum dissectione discunt, & vivis hominibus, bonâ valetudine utentibus, vitam denuò instaurare, & moribundus Spiritum, nè exeat, detinere. Et in hâc equidem praxi nulla unquam ætas fuit (sive diligentiam indefessam, sive abundantem Inventorum fælicitatem Spectes) nostrâ hac aut ditior aut fœcundior; cujus assidua Cura ac Labor improbus nova non pauca adinvenit, antiqua correxit, & hanc artem ita illustravit, ut jam ferè ad culmen perfectionis summum pervenisse videatur. Quod si de Anatomicis videndum esset. Nostrates plurimi extiterunt Heroes, qui ut sidera Cæli, Clarissima medicinæ Lumina, & veluti Dii Mortales in hâc Arte emicuerunt; Inter quos quantâ cum famâ Harvæius, Entius, Glissonius, Whartonus, Charltonus, Willisfius, Gualterus Nedhamus, Lowerus, cum multis aliis scripserunt, ut Libri & nomina ipsorum celeberrima nullis delenda sæculis testantur. Jam tandem ad vos, Viri Apollinei, mea se dirigit Oratio, sub quorum Patrocinio Lucubrationes hasce ausus sum divulgare. De vobis dicere fas est; quod quasi Medicorum Principes doctrinæ Jatricæ fines producitis, & Industriâ quotidianâ Artem condecoratis, unâ & augetis; adeò ut ex Collegio vestro, veluti solo fertili,

Inventa

Epistola Dedicatoria.

Inventa fælicissima ad artis propagationem, & Morborum extirpationem indies nascantur. Ad me quod attinet, Non ita arrogens sum, ut aliorum aut dicta, aut facta mihi ascribere velim, sed potius suum cuique tribuere cupiens, Authorum nomina ad marginem Libri appono, quos in Elencho (eos referente, é quorum scriptis universum hoc opus congestum fuit) Invenietis. Nostrum itaque, qualecunque Munusculum, etsi non verbis politum, attamen ad utilitatem hominum aliquo modo conducens, (quandoquidem veritatis plurimum in se contineat) fronte serenâ accipite. Valete & Æternum quàm fælicissimi vivite,

    Sic Orat,

        Nominib. Illustrib. & Nobillissim.

          Devotissimus & Observantissimus,

                Joh. Brown.

To his Highly Esteemed Friend Mr. John Browne, &c.

SIR,

I Have through want of leisure, but cursorily viewed your History of, and Observations on the Muscles: together with your choice number of their Types and Figures elegantly delineated with your own hand. Yet I think, I did it not so superficially, but that I may truly and confidently say, that whoever considers how profitable, and indeed how necessary the knowledge of the Muscles is, not only in Chirurgery, but in Medicine and Philosophy also, (since the faculty and use of the Nerves which distributes Life and Motion to all parts cannot be understood and explain'd without it) must, when he hath perused your Brief, Accurate, and Judicious Treatise, of the Muscles, together with your Observations and Animadversions thereon, be so Ingenuous, if he be just, as to acknowledge their great usefulness not only to Chirurgeons, but to Students in Medicine and Phylosophy likewise, and must therefore joyn with me to their Encomium; and earnestly desire, and importune their Publication as well as

Sir,

Your assured Friend and Servant,

E. Dickinson M. D.

Physician to the Kings Person and Family.

To Mr. John Browne on his Treatise of Muscles.

SIR,

I Cannot but commend your Ingenuous Design of making so useful a Book for the benefit of young Chirurgeons, and your Judicious choice of Cuts for that purpose: It is a Book which I am sure the Chirurgeons-Hall doth much want, for the younger sort to have in their hands in order to their better attending the Learned Reader. And not only so, but the most able Anatomists will be glad of so compleat a Contraction of so considerable part of Anatomy into one view, where they may at any time recollect themselves, as to their former Readings and Observations. Therefore your Edition of this Treatise cannot but oblige the World, and amongst others,

Charter-House,
April 14. 79.

Sir,

Your humble Servant,

Wal. Needham.

Myographicum hunc Tractatum ab Expertissimo Viro D. Johanne Browne conscriptum, Medicorum Chirurgorumque Tyronibus, & utrorumque etiam Peritissimis utilem fore Judicat,

Tho. Allen M. D.

Regius Medicus Oridinarius & Coll.
Lon. Soc. & Censor.

In Laudem Authoris.

PErlegendo vestrum de Musculis Libellum, magnum invenio certamen inter Ingenium & Ingeniosum tui calamum, quis Palmam portare meruit; & nè iniquè Judicarem, Legentibus sat erit, tam perutilem, sic perjucundum fore existimat,

Amicus tuus tui ad Aras,

Edvardus Warner.

Med. Personæ Regis Ordinarius
& ejus Exercitui M. Generalis.

To his much valued Friend Mr. John Browne.

S I R,

IHave read over your Treatise of Muscles, (as my time would permit) and find you have taken great pains and care in your Collections, and made good Observations, and exprest much Art and Industry in your Figures: and (by consulting excellent Authors) have contracted much into little: that a competent knowledge in this part of Anatomy (so absolutely necessary in the safe practice of Chirurgery) may probably be much easier attained to, than formerly by young Indagators; and be a further confirmation to others, who have already sweat in the same Study. Therefore I cannot but encourage you to make it publick, by which I suppose you will oblige many who are Friends to such Anatomical Enquiri:s: amongst whom give me leave to reckon,

Sir,

Martii 26. 79.

*Your Affectionate Friend
To serve you,*

Edmund King M.D.

Physician in Ordinary to his Majesty.

TO

TO THE
Ingenuous and Studious
READER.

Curteous Reader,

S *Anatomy is generally allowed the beſt and trueſt Speculum or Looking-Glaſs, illuſtrating, and preſenting all the parts of the Body, with their Affects, and Affected Parts, moſt clearly and evidently: ſo alſo is it the main Baſis, on which Chirurgery doth build its faireſt Fabrick, and comely Structure. And of what great uſe this hath been, and is, to Mankind, not only the Ancients, but alſo Modern Writers do moſt admirably declare; How this Renowned Art hath in former Ages been treated, and carefully treaſured up and Celebrated in Theaters and Anatomical Schools, many Authors have to Poſterity voluminouſly ſatisfied; And that it ſtill doth get greater footing amongſt the Ingenuous and Induſtrious Sons of Art, every Age doth, and may with admiration evince, and the many Learned Lectures annually read, may both moſt honourably and happily ſet forth and demonſtrate.*

This Art of Anatomy, as it doth require the induſtrious Study, and accurate care of a curious Enquiry, ſo alſo doth it reward his pains with the Jewel of Knowledge, and afford his Endeavours the happy iſſue of a fair ſucceſs: giving thoſe methods and meaſures of ſecurity

The Epistle to the Reader.

curity in Chirurgical Practice, which are most consentaneous and genuine for the *Industrious Practitioner*, to *study* and *search after*. This is as his best *Compass*, whereby he may *sail freely* in the main Ocean of his *Art*, without any *mistrust* or *dread* of *splitting* himself upon any *Rocks* of *danger*, or *Shelves* of *mistake*, for its by this alone, that we arrive at and do gain how our *Membranes* are made, and *spun* out of *Spermatick Fibres*, and how out of *Maternal Blood* are framed our *Muscular parts*. It is from this *Tree of Knowledge*, all the differences between a *Vein* and *Artery* are *pluckt up*: Its this *Garden* that affords us the various *Flowers* of *Nerves* and *Tendons*, and shews us the different *Beds* of *Ligaments* and *Bones*: It's this *Sun* that conducts the *Understanding*, that advanceth the same into *Affection*, that promotes our *Affections* towards our *Maker*. And come we but to take a view and survey of the *symmetry* of his parts, and therein but cast our *Eye* and consider on the rare *Offices* and *Uses* which are variously bestowed on them, we ought as well in *Justice*, as in *Reason*, to grant and allow, that nothing in this great *Universe* may or can surpass him, especially in the exquisite *Fabrick* of his *Muscles*, and the variety of their *Motion*.

Now although *Myology* hath been by the *Ancients* thus traced, yet *Steno* hath quite altered the *Fabrick* of their design, affirming, that there is in every *Muscle* two opposite *Tendons*, into which are inserted two kinds of *Fibres*, the which being closely annexed do frame the *Tendon*, and in that part wherein they are loosely intertext, there they do allow of a constituted *Flesh*, implanting one above another, here placing and disposing the thickness and depth of the *Flesh*, framing its latitude, and declaring its order of *Fibres*; and this *Figure* he describes by a *Mathematical Line*, taking thence

their

The Epistle to the Reader.

their *Cannons* which do explain their actions; *Thus he satisfies, that all the Fleshy Fibres in a simple Muscle, are carried in a direct Line from one Tendon obliquely into another, and these Tendons are interwoven in their opposite ends and angles of the Fleshes*; by which he

Steno's description of a Muscle.

doth ingeniously describe a *Muscle* to be a collection of moving Fibres, so framed and formed, that the middle Fleshes do frame an oblique angle, and that the two opposite Tendons do form two *Quadrangular Prisma's*; this Figure he very accurately represents by *Instruments* used by *Painters*, or *Picture-Drawers*, wherein we see by the application of their *Pegs*, the true insertion of their *Tendons* are hereby very well explained, and the Fleshy parts lively delineated: for when they bring their greatest distances from their opposite angles, and being hereby made more acute, these are brought to two sides, and so this *Instrument* is made both longer and narrower, this shewing the *Muscle* no way contracted, but rather narrowed, and reduced into a longer body, the length thereof not being changed, save only in its *Position*; Thus have we by Steno shewn, a muscle may appear as a *simple* part consisting of one Venter and two *Tendons*, as are many of those in the *Arm*, or *Thigh*, and in many other parts of the *Body*; or *Compound* having many Venters, to every of which is allowed two *Tendons*.

In a *simple Muscle*, because either one *Tendon* or both ought to be attracted, and because the attracting part is thin, and broad, the *Belly of the Muscle* doth require for its frame, a diversity of *Fibres* to answer to the variety of its *Figures*; whereas *Compound Muscles* are distinguished by their greater variety of *Fibres*: for besides this variety of *Figures* which ought to be observed; so also ought their *Venters*, being either more or less, with their order of *Fibres*, and diversity

of

The Epistle to the Reader.

of Frames *have a respect allowed them.* Therefore *shall we find that to every simple, as well as compound* Muscle *is bequeathed a* Membranous *covering, invested* [Membrane] *with Fleshy Fibres for its direct motion, and is intertext with transverse Fibres for forming it into a Membrane. It hath a Vein and Artery inserted into its middle, send-* [Vein and Artery] *ing forth of their Surcules into the oblique moving Fibres, from whence the smaller* Ramifications *being dispersed through the* Interstitia *of the Fleshes, doth besprinkle and bedew them with their Afflux of Blood,* [Nerve] *and its* Reflux : *The* Nerve *entring herein doth distribute of its small Branches into its neighbouring Muscles, carrying in it the soul of the commanded action, and commanding its execution in the lower guard of the Fleshy Fibres, and of its Membranous Fibrilla's ; And as touching its action, in dissection of living Bodies, we* [Its action] *plainly perceive that it doth contract, but not as the old opinion held : that the Fibres did contract from their ends, towards their Originations, one end of the Muscle being carried through the other ; whereas the Fleshy Fibres only, and their ends are seen to be contracted towards their middle, their Tendons being mutable, and not altered either in their longitude or thickness, the which Worthy* Steno *first observed, the which he doth describe in the* Diaphragma, *and those* Muscles *appointed for* Respiration, *the which is moved by a constant turn, as is cleared and perceived ; How oft therefore the Muscle is seen to be contracted, all its Fleshy Fibres in either end are apparently shewn to be driven together, and as it were seen to bow and yield to each other ; and hence do seem to appear either shorter or thicker : the which being loosned from its constriction, you will find it to appear in its proper length and thickness, this alteration being produced by the Spirit or subtile* Matter *which passeth from the Tendons into the Fleshy parts ;*

and

The Epistle to the Reader.

Tendinous Fibres, and do at length get quarter there, and entertainment, as in other Promptuaries or Mansions, the which Spirits being in their nature very active, so fast as their vigour will permit them, they do expand themselves, and penetrate into the Fleshy Fibres: and continuing this their course, they at length do arrive at the Tendons, and having once entred them, they do therein proceed in the same method: and that the Animal Spirits flowing from the Tendinous Fibres, do equally pass under the Fleshy Fibres, is very apparent; in that it is granted that in every Muscle there is allowed two Tendons, whose opposite Angles are so framed, that these Animal Spirits running from a double top do fill the whole Body of the Muscle; and the motion hence taking its Origination, doth very speedily receed, if the contraction ought to be made indifferently towards the middle Fleshy parts: the Tendons are generally equal, but the motion most inclinable towards one part of the Flesh: and hence may we collect, the regular or irregular motions of a Muscle, every regular or irregular motion hereof granted to arise from either the Cerebrum *or* Cerebellum, *it being thence dispatcht by the Nerves, and so sent into the Muscles, the effects and consequences whereof do evidently evince and demonstrate. Thus have I a little presumed to enlarge upon the Readers Epistle, as touching the use and benefit of the Muscles, wherein also I have introduced somewhat of Muscular Motion according to* Steno, *and have the rather chosen to enlarge the Discourse here, the Body of the Treatise being wholly intended to appear Publick and Concise, without any Flourishes or empty Enlargements of Discourses or Controversies.*

Peruse

The Epistle to the Reader.

Peruse therefore this ensuing Discourse with a Candid Interpretation, and pass by all the Literal Elapses you may meet with, and accept these with that Amicable Mind, as they are Dedicated and intended by,

John Browne.

From my House at the
  Chirurgeons-Arms at
  Charing-Cross,
London.

The Names of the Subscribers.

A.

	l.	s.	d.
Christopher Duke of Albemarle	02	03	00
Henry Earl of Arlington Lord Chamberlain of His Majesties Houshold	01	01	06
Henry Earl of Arundel	02	03	00
Henry Earl of St. Albans	01	01	06
Robert Earl of Alisbury	00	10	00
Thomas Allen M. D.	00	10	00
Adam Angus M. A.	01	00	00
Richard Adams M. B.	00	10	00
George Aldebar M. A.	00	10	00
Thomas Allen Gent.	01	00	00
John Anderson Chir.	00	10	00

B.

William Earl of Bedford	01—01—06
John Earl of Bath	01—01—06
Charles Baueclear Earl of Burford	05—00—00
George Lord Berkley	00—10—00
Sir John Baber	01—01—06
Sir Nicholas Bacon	01—00—00
Sir Edward Baesh	0—01—06
William Bell D. D. His Majesties Chaplain	00—10—00
John Butler D D. Prebend of Windsor	00—10—00
Francis Bridge D. D. His Majesties Chaplain	00—10—00
Robert Brady M. D and Regius Professor of Physick at Cambridge for himself and Caius-Colledge	01—00—00
Samuel Blyth D. D. for Clare-Hall	00—10—00
Peirce Brackenbury M. D.	00—10—00
S. muel Beck M. A.	00—10—00
John Batly M. A.	00—10—00
Thomas Bambrig M. A.	00—10—00
Joshua Barnes M. A.	00—10—00
Arthur Bury D. D. at Oxon	00—10—00
John Bainbrigg Gent. at Oxon	00—10—00
Proster Balch of Wadham-Colledge	00—10—00
Sir Tho. Browne of Norwich M. D.	01—00—00
Peter Parwick M. D.	00—10—00
Edward Browne M. D.	00—10—00
William Priggs M. D.	00—10—00
Robert Foyle Esq	01—00—00
William Bridgman Esq.	01—00—00
James Beverly Esq;	00—10—00
Henry Bedingfield Esq;	00—10—00
James Bagnal Esq;	00—10—00
John Brown Cler. Parliamentor. Esq;	01—00—00
Philip Browne Gent.	00—10—00

C.

William Lord Archbishop of Canterbury	01—00—00
Henry Earl of Clarendon	01—01—06
John Lord Bishop of Chester	00—10—00
Tho. Lord Cromwel Earl of Ardglass	01—01—06
Benjamin Calamy D. D. His Majesties Chaplain	00—10—00
John Clerke D. D. for himself and Colledge	01—00—00
Ralph Cudworth D. D. for Corpus Christi-Colledge	00—10—00

	l.	s.	d.
William Cooke D. D. for himself and Jesus-Colledge	01	00	00
Thomas Coxe M. D.	01	00	00
Samuel Collins M. D.	00	10	00
Andrew Clench M. D.	00	10	00
John Clerke M. D.	00	10	00
Hugh Chamberlain M. D.	00	10	00
Richard Colinge Esq;	01	00	00
John Cooke Esq;	01	00	00
John Cresset Esq;	00	10	00
William Chapman Esq;	00	10	00
Thomas Coxe Esq;	01	00	00
Benjamin Colinge Gent.	01	00	00
Charles Chapman Gent.	00	10	00
James Cooke Gent.	01	00	00
John Clerke Gent.	00	10	00
Nathaniel Coxe Gent.	00	10	00

D.

William Earl of Denbigh	00—10—00
Tho. Earl of Danby	01—00—00
Sir Edward Deering	00—10—00
Edward Dickinson M. D. Physician to His Majesties Houshold.	01—01—06
John Downs M. D.	00—10—00
Robert Davy Esq;	00—10—00
Richard Dalton Esq;	00—10—00
Thomas Dunckley Gent	00—10—00
Peter Dearines Gent.	00—10—00
Walter Drury Apothecary	01—00—00
George Deare Apothecary	00—10—00
Peter Dent M. B.	00—10—00

E.

Arthur Earl of Essex	01—01—06
Tho. Lord Howard of Eschrick	00—10—00
Peter Lord Bishop of Ely	00—10—00
Peter Elliot M. D at Oxon	00—10—00
John Eachard D. D. for himself and Catherine-Hall in Cambridge	01—00—00
Samuel Elmore Chir.	01—01—06

F.

Lewis Lord Duras Earl of Feversham	01—01—06
Robert Lord Ferrers	00—10—00
Ralph Flyer M. D. at Cambridge	00—10—00
Sir Alexander Frasier M. D. Def.	00—10—00
Phineas Fowke M. D.	00—10—00
Thomas Fetherstonhalg Esq;	00—10—00
Martine Folke Esq;	00—10—00
James Frafer Gent.	00—10—00
Tho. Feild Gent.	00—10—00
John Francklin Chir.	00—10—00
John Fage Gent.	00—10—00

G.

Henry Duke of Grafton	02—03—00
John Goad D. D.	00—10—00
William Gibbons M. D. at Oxon	01—00—00
Humphrey Gower D. D. Vice-Chancellor of Cambridge, for himself and St. John's-Colledge	01—00—00
John Gosslin M. D. at Cambridge	00—10—00
Charles Goodall M. D.	00—10—00

Christopher

The Subscribers Names.

Name	l.	s.	d.
Christopher Green M. B.	00	10	00
William Gold M. A.	00	10	00
John Gadbury Student in Astrology	00	10	00
Peter Gilsthorp Apothecary	00	10	00
Allen Gyles Apothecary	00	10	00
Richard Green Bookseller, for six Books in Quires.	01	05	00

H.

Name	l.	s.	d.
George Viscount Hallifax	01	01	06
Sir Philip Howard	00	10	00
Sir John Hobart	01	01	06
Sir Michael Hickes	00	10	00
William Holder D. D. Subdean to His Majesties Chappel	00	10	00
Dr. Hascard D. D. Chaplain to His Majesty	03	10	00
Tho. Holbeck D. D. for Emanuel-Colledge Camb.	00	10	00
Edward Hulse M. D.	00	10	00
Peter Hume Gent.	00	10	00
Tho. Hollyer Chirurgeon of His Majesties Hospitals	00	10	00
John Hollyer Gent. for two Books	01	00	00
Tho. Harper Chir.	00	10	00
Henry Hern Apothecary	00	10	00

I.

Name	l.	s.	d.
Sir William Jennings	00	10	00
William Jane D. D. and Regius Professor of Divinity at Oxon	00	10	00
Gilbert Ironsides D. D. President	00	10	00
Charles James M. A.	01	00	00
Henry James D. D. for Queens-Colledge at Cambridge	00	10	00
James Jackson M. D. at Camb.	00	10	00
Tho. Jamson Esq;	00	10	00
Gabriel Jones Chir.	00	10	00

K.

Name	l.	s.	d.
Anthony Earl of Kent	00	10	00
Sir John Kirke	01	00	00
Edmund King M. D. Physician in Ordinary to His Majesty	01	00	00
John Knight M. D. Principal Chir. to his Majesty def.	01	00	00

L.

Name	l.	s.	d.
Henry Lord Bishop of London	00	10	00
Henry Lord Bishop of Lincoln	00	10	00
Aunger Lord Longford	00	10	00
Sir Peter Lely def.	01	00	00
John Lamphire M. D. History Professor at Oxon	00	10	00
John Luff M. D. Regius Professor of Physick at Oxon	00	10	00
William Levenz M. D. President of St. John's-Colledge, Oxon, for himself and Colledge	01	00	00
John Ludwell M. D. at Oxon.	00	10	00
Richard Lydall M. D. at Oxon.	00	10	00
John Lawson M. D.	00	10	00
Christopher Ludkin M. B.	00	10	00
John Leeger Chir.	00	10	00
Tho. Langham Apoth.	00	10	00

M.

Name	l.	s.	d.
James Duke of Monmouth	01	03	00
Tho. Lord Morely and Mounteagle	01	01	06
John Montague D. D. Clerk to His Majesties Closet.	00	10	00
Tho. Marthall D. D. His Majesties Chaplain	00	10	00
Henry More D. D. at Cambridge	00	10	00
Sir John Micklethwait M. D. President of the Colledge in London	01	00	00
Sir Thomas Millington M. D.	00	10	00
Ferdinand Mendez M. D. Physician to the Queen	00	10	00
George More Esq;	01	00	00
John Malyverer M. A.	00	10	00
Richard Mills Chir. R.	00	10	00
Martine Mey Chir.	00	10	00
Nicholas Mosely Apoth.	00	10	00

N.

Name	l.	s.	d.
Henry Duke of Newcastle	01	01	06
George Earl of Northumberland	01	03	00
James Lord Norries	01	01	06
Walter Needham M. D.	01	00	00
Tho. Newcombe Junior	01	00	00
John Northleigh Gent.	00	10	00

O.

Name	l.	s.	d.
Aubery Earl of Oxford	01	01	06
John Lord Bishop of Oxford	00	10	00
George Oliver M. A.	00	10	00

P.

Name	l.	s.	d.
William Lord Paston	01	01	06
Sir John Pettus	01	01	06
Simon Patrick D. D. His Majesties Chaplain	00	10	00
John Price D. D.	00	10	00
Robert Pepper D. L. L. Chancellor of Norwich	00	10	00
Sir Tho. Page Provost of King's-Colledge in Cambridge	00	10	00
John Peachel D. D. for himself and Magdalen-Colledge, Camb.	01	00	00
For Pembroke-Hall	00	10	00
Henry Paman M. D.	00	10	00
Robert Pitt M. D. Anatomy Professor	00	10	00
Jos. Pullein S. T. B.	00	10	00
John Packer M. B.	00	10	00
George Payne of Clare-Hall Oxon	00	10	00
Robert Paston Esq;	00	15	00
Roger Pope Esq;	00	10	00
William Prince Esq;	00	10	00
George Perin Esq;	00	10	00
Robert Power Gent.	00	10	00
James Pearse Esq; Chirurgeon to the King's Person	01	00	00
William Pearse Chirurgeon to His Majesties Hospitals	00	10	00
John Partridge Student in Astrology	00	10	00
Tho. Prescott Apothecary	00	10	00

Q.

Name	l.	s.	d.
Gabriel Quadring M. A.	00	10	00

R.

Name	l.	s.	d.
His Highness Prince Rupert	05	00	00
Charles Duke of Richmond	01	03	00
Thomas Earl Rivers	01	01	06
John Ratcliff M. D.	00	10	00
John Ruddlon B. L. L.	00	10	00
William Rowley Gent.	00	10	00

The Subscribers Names.

	l.	s.	d.
William Rowe *Apothecary*	01	00	00
William Rapier *Apoth.*	00	10	00

S.

	l.	s.	d.
Tho. *Earl of* Sunderland	01	01	06
Anthony *Earl of* Shaftsbery	01	00	00
Seth *Lord Bishop of* Salisbury	00	10	00
George Stradling *D. D. His Majesties Chaplain*	00	10	00
Tho. Spratt *D. D. His Majesties Chaplain*	00	10	00
John Sewmears *D. D. Dean of* Gernesy	00	10	00
Gregory Scott *D. D.*	00	10	00
Sir Tho. Sclater *M. D.*	00	10	00
For Sydney-Colledge Camb.	00	0	00
John Spencer *D. D. for* Corpus-Christi-Colledge Camb.	00	10	00
Sir Charles Scarborough *M. D. Principal Physitian to the King*	00	10	00
Nicholas Staphurst *M. B.*	00	10	00
Francis Smith *M. A.*	00	10	00
Mr. Sagittary *M. A.*	00	10	00
Bevill Skelton *Esq;*	00	10	00
Henry Street *Gent.*	01	00	00
Edward Syston *Gent.*	00	10	00
Edward Snape *Gent.*	00	10	00
Samuel Staynes *Gent.*	01	00	00
Tho. Sydny *Gent.*	00	10	00
Daniel Sneaton *Chir.*	00	10	00
Henry Staff *Chir.*	00	10	00
Zachariah Skillcarn *Chir.*	00	10	00

T.

	l.	s.	d.
Sir Richard Tufton	00	10	00
Tho. Tenison *D. D. His Majesties Chaplain*	00	10	00
Dr. Thistlethwait *D. D. His Majesties Chaplain*	00	10	00
George Thorp *D. D.*	00	10	00
Edward Tyson *M. D.*	00	10	00
Samuel Tryon *Esq;*	00	15	00

	l.	s.	d.
John Topham *Esq;*	01	00	00
William Tovey *Gent.*	01	00	00
Edmund Themylthorp *Gent.*	00	10	00
Mr. Tyndall *M. B.*	00	10	00
Edmund Theorold *Chir. Master of the Company*	01	00	00

V.

	l.	s.	d.
Philip Vendosme *Grand Prior of* France	01	03	00

W.

	l.	s.	d.
Henry Lord *Marquess of* Worcester	01	01	06
Edward Lord Ward	00	10	00
Dr. Wallis *D. D. Geometry Professor at* Oxon	00	10	00
Thomas Witherly *M. D. and Physician to the Kings Person*	00	10	00
Daniel Whystler *M. D.*	00	10	00
Edward Warner *M. D. Physitian to the King*	00	10	00
Robert Werden *Esq;*	01	00	00
John Wynyard *Esq.*	01	00	00
John West *Esq;*	00	10	00
William Williams *Esq;*	00	10	00
John Walker *Junior Gent.*	01	00	00
Powel Williams *Gent.*	00	10	00
Hugh Willoughby *M. A.*	00	10	00
Owen Wynne *Gent.*	00	10	00
Richard Warr *Gent.*	00	10	00
Sackvil Whittle *Reg. Chir. def.*	01	00	00
Jaques Wiseman *Chir.*	00	10	00
Doughty Wormell *Chir.*	0	10	00
Josias Westwood *Chir.*	00	10	00

Y.

	l.	s.	d.
Thomas Yates *Gent. President of* Brason-Nose Colledge *Oxon.*	00	10	00
Robert Yard *Gent.*	00	10	00
John Young *Gent.*	00	10	00

This

This Table sheweth the Names of the Muscles, as they do arise in Dissection.

Obliquus Descendens
Obliquus Ascendens
Rectus
Piramidilis
Transversus
Frontalis
Aperiens Palpebrani Rectus
Claudens oculum superior
Claudens oculum inferior
Recti Quatuor Oculi
Obliquus Primus Oculi
Obliquus Secundus Oculi
Attollens Aurem
Detrahens Aurem
Adducens Aurem
Abducens Aurem
Externus Tympani Auris
Internus Tympani Auris.
Abducens Nasi alas
Attollens Nasi alas
Claudens nasum externus
Claudens nasum internus
Communis Claudens Alas
Zygomaticus Riolani
Abducens Labios
Deprimens Labii inferius
Constringens Labios
Platysma Myodes sive Quadratus
Buccinator
Masseter sive Mansorius
Temporalis seu Crotaphites
Mastoideus
Biventer sive Digastricus
Coracohyoideus
Sternohyoideus
Sternothyroideus
Hyothyroideus

Styloceratohyoideus
Pterygopalatinus
Spheno-palatinus
Mylohyoideus Riolani
Geniohyoideus
Miloglossus
Ceratoglossus
Genioglossus
Hypsiloglossus
Styloglossus
Lingualis
Cricothyroideus anticus
Æsophigæus seu Sphincter Gulæ
Stylopharyngæus
Cephalopharyngæus
Cricoarytenoideus Posticus
Cricoarytenoideus Lateralis
Arytenoideus
Thyroarytænoidus
Sphænopharyngæus Primus
Sphænopharyngæus Secundus
Pterygoideus Externus
Pterygoideus Internus
Longus
Scalenus seu Triangularis
Pectoralis
Subclavius
Serratus major Anticus
Serratus minor Anticus
Intercostales Externi
Intercostales Interni
Cremasteres
Erector sive Director Penis
Accelerator Penis
Musculi Clitoridis
Levatores Ani
Sphincter Ani

Sphincter

The TABLE.

Sphincter Vesicæ
Detrusor Urinæ
Diaphragma.

Here let the Body be turned upon the Face.

Cucullaris sive Trapezius
Latissimus Dorsi
Rhomboides
Levator Patientiæ
Rotundus Major
Superscapularis Superior
Superscapularis Inferior
Nonus humeri Placentini sive Rotundus Minor
Subscapularis

If you intend to take off the whole Arm with the *Scapula*, the Dissection of these following Muscles will with more ease be performed.

Deltois
Biceps
Octavus humeri Placentini sive Coracobrachialis
Brachiæus Internus
Gemellus Major
Gemellus Minor
Anconæus
Palmaris
Caro Musculosa Quadrata
Flexor Carpi Interior seu Ulnaris
Flexor Carpi Exterior sive Radialis
Flexor Secundi Internodii Perforatus
Flexor Tertii Internodii Perforans
Flexor Tertii Internodii Pollicis
Pronator Radii Teres

Pronator Quadratus
Flexores Primi Internodii Digitorum
Flexor Primus, Primi Internodii Pollicis
Flexor ejusdem Secundus
Flexor Secundi Internodii Pollicis
         Primus
         Secundus
         Tertius
         Quartus
Minimi Digiti abductor
Pollicis abductor
Pollicis adductor
Interossei
Extensor Carpi Exterior
Extensor Secundi & Tertii Internodii Digitorum
Supinator Radii Longus
Extensor Pollicis ossis Tertii
Extensor Secundi & Tertii Pollicis.
Abducens Indicem
Supinator Radii Brevis
Primi Internodii Extensores.

Here you return to the Body it self as it lies

Serratus Posticus Superior
Serratus Posticus Inferior
Splenius sive Triangularis.
Trigeminus
Transversalis
Spinatus
Recti Majores
Recti Minores
Obliqui Superiores
Obliqui Inferiores
Longissimus Dorsi
Sacrolumbus
Cervicalis descendens
Sacer
        Semispinatus

The TABLE.

Semispinatus
Quadratus
Psoas
Psoas Parvus

If you please to take off the Thigh from the Trunck of the Body, by dividing the *Os Ileon* from the *Os Sacrum*, the dissection of the subsequent Muscles will the better be performed.

Iliacus Internus
Glutæus Major
Glutæus Minor
Glutæus Medius
Piriformis sive Iliacus Externus
Obturator Internus
Quadrigeminus
Obtuator Externus
Membranosus
Sartorius
Gracilis
Rectus
Vastus Externus
Vastus Internus

Biceps
Semimembranosus
Seminervosus
Triceps
Lividus
Gasterocnemius Externus
Plantaris
Gasterocnemius Internus
Subpopliteus
Flexor Tertii Internodii Perforans.
Tibiæus posticus
Flexor pollicis
Flexor Secundi Internodii Perforatus.
Adducens Pollicem
Abducens minimum Digitorum
Transversalis Placentini
Tibiæus Anticus
Peroneus Primus
Peroneus Secundus
Extensor Pollicis
Extensor Tertii Internodii Digitorum
Extensor Secundi Internodii Digitorum
Interossei.

The Names of the Authors concerned in this Muscular Discourse.

Thomas Bartholinus
  Casper Bauhinus
Albertus Columbus
Bartholomeus Cabroules
Isbrandus Diemerbroeck
Galen
Regnerus de Graaf
Hipp crates
Andreas Laurentius

Julius Casserius Placentinus
Johannes Riolanus
Daniel Sennertus
Adrianus Spigelius
Joannes Valverdus
Andreas Vesalius
Joannes Veslingius
Vidus Vidii

Obliquus

Obliquus Descendens.

THis first pair of Muscles with which Nature hath covered the Abdomen, as a Vail, are endowed with oblique Fibres, and by reason of their descent, they have this name bestowed upon them. They do arise from the lower parts of the 6th. 7th. 8th. and 9th. Ribs indented; or indenting themselves with *Serratus Major Anticus*; as also Membranous from the transverse processes of the Vertebres of the Loyns, and part of the *Os Ileon*, and then passing to the *Linea Alba*, and *Os Pubis*, by a broad Nervous Tendon marches into the middle of the Abdomen; (its very hard to separate this Muscle from its subjacent Tendon without laceration) it adhering so closely to it. This Tendon with its next neighbour being either lacerated or dilated, and the *Omentum* or *Intestines* hereby making a prolapsion either into the *Inguen* or *Scrotum*, does occasion either an *Hernia Omentalis*, or *Intestinalis*. To dissect this Muscle exactly, you must divide the *Latissimus* from him very low, so as that you may attain his Lumbal Origination the better.

<small>*This doubly com-*</small>
<small>*terally com-*</small>
<small>*pressi the Ab-*</small>
<small>*domen.*</small>

<small>*Obs.*</small>

The chief use of this Muscle, as *Columbus* and *Laurentius* do affirm, is to contract the *Thorax*, as sometimes upwards, hereby assisting Respiration. *Laurentius*'s observation of these Muscles is worth note; when he writes that these Abdominal Muscles are quite contrary to the other Muscles of the Body, these being crooked before they do Operate, and do turn inward in their Operations, hereby with much facility compressing the inward Cavities; the which, by reason of the laxness of the lower Belly, and its yielding to vacuity, it frames a Contention inwards, and a Remission outwards. Another Observation may be, that these Muscles do bind the Intestines in oblique descending Angles.

<small>*Its use.*</small>

<small>*Obs.*</small>

The Explanation of the First Table.

B *Part of the Muscle called* Ani-scalptor.
D. Musculus Pectoralis in situ.
C. Serratus Major Anticus.
G.G. I.I. K.K.K. Musculus obliquus descendens.
G.G. *Shews the Fleshy part of this Muscle.*
a.a.a.a. *Shews its Connexion with* Serratus Major.
I.I. *Shews another part of the same Muscle.*
K.K.K. *Shews another Fleshy part of this Muscle, with its Tendon and where it begins, and how it is expanded into the* Linea alba.
V.V. *Shews the Semilunary Line.*
L.L.I. *The Tendon of the Oblique descendent Muscle running over the right Muscles of the Abdomen to the* Linea alba.
M.M.M. *The White Line into which this Tendon is inserted.*
N.N.N. *The Intersections of the Right Muscles.*
P.P. *The Spine of the* Os Ileon.
Q.Q. *Some heads of the Muscles moving the Thigh.*

Obliquus

TAB: I.

Obliquus Ascendens.

THis is immediately substrated to the former, and is furnished with Ascendent Fibres; it ariseth fleshy from the Appendix of the *Os Ileon,* and Membranous from the processes of the same Vertebres as the former; and so ascending into the 11*th.* and 12*th.* Ribs with a fleshy Margent, then extending himself into a large double Tendon in which the *Rectus* is conveyed, proceeds to the *Linea alba,* and Semilunary Line, and is implanted into the 9*th.* 10*th.* 11*th.* and 12*th.* Ribs. Observe that this Muscle is best rais'd by finding the Nerve that runs between this and the Transverse at his Origination from *Os Ileon.* *This doth help the former in its Compression.*

Obs.

The chief use and action of this Muscle as *Columbus* doth offer, is, That these working together, they do detract the Muscles of the *Thorax.* Nature planting here a Series of Fibres contrary to the former, and these being by her thus made for a stronger Compression. *Its use.*

Moreover as I humbly conceive, that this Oblique ascending Muscle is obliquely perforated near the *Os Pubis,* by the Cremasters and the Spermatick Veins and Arteries a little above the perforation of the Oblique descending Muscles; so that these Oblique ascending Muscles lodging just under the former, do run counter with them in their Fibres, and do keep them in Oblique ascending Angles. *obs. 17.*

Moreover, that whereas the various Muscles of the Abdomen several ways contracting themselves inwards, do force the Excrement downwards, and at the same time do reduce both the Ventricle and Intestines into their proper places, and are Antagonists to the *Diaphragma,* because in its motion in order to enlarge the capacity of the *Thorax* to give reception to the Lungs tumefied with Air, the *Diaphragma* is brought towards a plain, and doth thereby both press the Stomach and Intestines downwards, by which, the motion of the Chyle is gently sollicited into the *Ductus chiliferus.* And when the *Diaphragma* hath done playing, these Abdominal Muscles do act their parts, by relaxing the *Diaphragma,* and bringing it into an Arch, the Belly growing lank, and the Intestines and Stomach being forc't inwards, and upwards, by the contraction of the Abdominal Muscles.

The

The Explanation of the Second Table.

A *A.* Obliquus Descendens, *laid bare.*
B.B.B. The bodies of the Ribs.
9.10.11. *The lower Ribs.*
a a.a. Shews the Tendinous Membrane of the Oblique descendent Muscle.
B.B. Obliquus ascendens, in situ, *shewing its Ascendent Fibres.*
C.C.C. Linea Semilunaris.
i.i.i.i. Recti Musculi, *Transparent under the Tendons of the Oblique ascendent Muscle.*

Rectus.

TAB. II.

Rectus.

THis third pair are cloathed with Right Fibres, being *This Muscle brings the Belly forwards.* made very ftrong, and well lin'd with Flefh: They do arife from the *Os Pubis*, and running according to the length of the Body, are inferted into the fides of the *Sternon*, where the laft true Ribs have their Cartilages. The Infertions hereof are various, for fometimes there is feen three, fometimes four, and fometimes three and a half; fometimes all above, fometimes fome below; as alfo the Anaftomofis of the Mamillary Veffels external, and Epigaftrick internal, always found in Women, rarely in Men; as alfo the multitude of Nerves fent to the Perigraphs is very obfervable, for if you find four Perigraphs, you will find no Pyramidal Mufcles.

Thefe Mufcles are allowed to be of great ufe; fome Authors *Ufe.* affirming that they do abduce the *Penis* from the Ribs. And when we do arife out of our Beds, thefe Mufcles do feem to tumefie and fill outwards; others do declare, that thefe do inflex the *Thorax*, and do draw the Breft to the *Os Pubis*, and the *Os Pubis* to the *Thorax*, and that out of their Contraction, there is feen two various Motions performed.

But I humbly conceive that thefe Mufcles taking their Origination from the *Os Pubis* and *Sternon*, and inferting themfelves into the *Linea alba*, and running all down in length through the middle of the Abdomen, do in their Contraction prefs the Infertions inwards, and do affift the Periftaltick motion in the exclufion of groffer Excrements.

B   The

The Explanation of the Third Table.

A. *The Cutis with the Fat laid bare.*
D.D.D.D. *The Right Muscles of the Abdomen.*
e.e.e.e. *The Intersections of these Muscles.*
i i. *The Pyramidal Muscles.*
L.L. *The Oblique Ascendant Muscles* in site.
M.M. *That part of the Tendon of the Oblique Ascendant Muscle which covers the Right Muscle.*
P.P. *The Intercostal Muscles.*

Pyramidalis

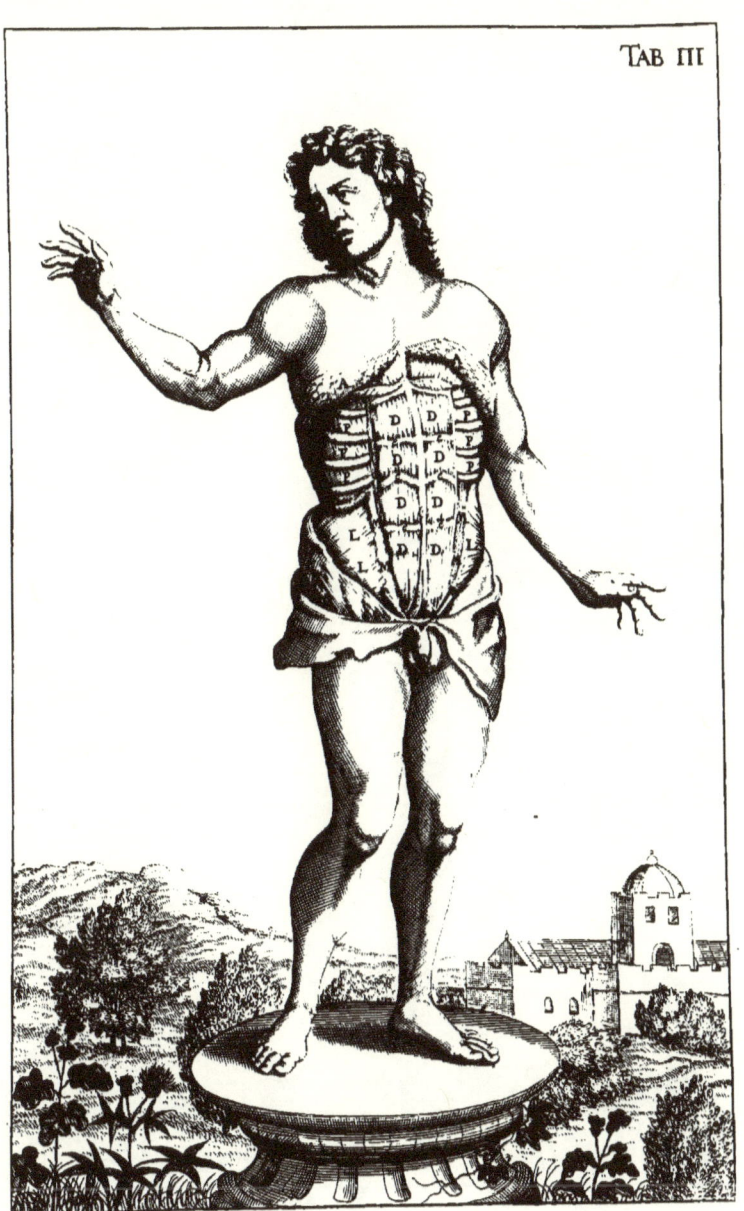

TAB III

Pyramidalis or Succenturiatus.

THis Muscle hath its name from its make, and resem- *This presseth the Abdomen downwards.* blance it carries with a Pyramidal Figure. It ariseth from the external part of the *Os Pubis* broad and fleshy, and running somewhat below the *Rectus*, is inserted by a long and round Tendon into the *Linea alba*. These Muscles are seen sometimes to be wanting in such Persons as have the Origination of the Ascendent Muscle not from the *Ileon*, but from the strong Ligament which runneth from the Spine up to the *Os Pubis*, internally: and have four Perigraphs in the *Rectus*.

Vesalius, Adernaus, and *Columbus* do all describe them (though badly) to arise from the beginning of the Right Muscles; But that these are distinct Muscles are evidently apparent.

Fallopius the Inventer of them, doth ascribe the action of *Their Use.* Compression to them; and that they do promote the Excretion of Urine, this also he doth affirm. *Laurentius* does observe, that if one of these Muscles doth work alone, it draws the *Linea alba* obliquely downwards; if they do work together, they do work it directly downwards, and do hereby compress part of the Inguen and the Bladder, when we at leisure do discharge our Urine. *Columbus* will have these Muscles somewhat to add to the Erection of the Penis, but his Opinion is much contradicted by *Flud* from their Situation: for they cannot reasonably be allowed to serve for this use, because they do in no measure reach this part, and are also apparent in Women.

The

The Explanation of the Fourth Table.

A *Shews the Muscle* Obliquus Ascendens
B B. *Demonstrates its Tendon.*
C.C.C.C. *The Right Muscle of the Left Side, shewing the Tendons Duplicature.*
K.K.K.K. *The same Muscle of the Right Side.*
d. *The Transvers: Muscle of the Abdomen.*
e.e. *The Pyramidal Muscles.*

Transversus.

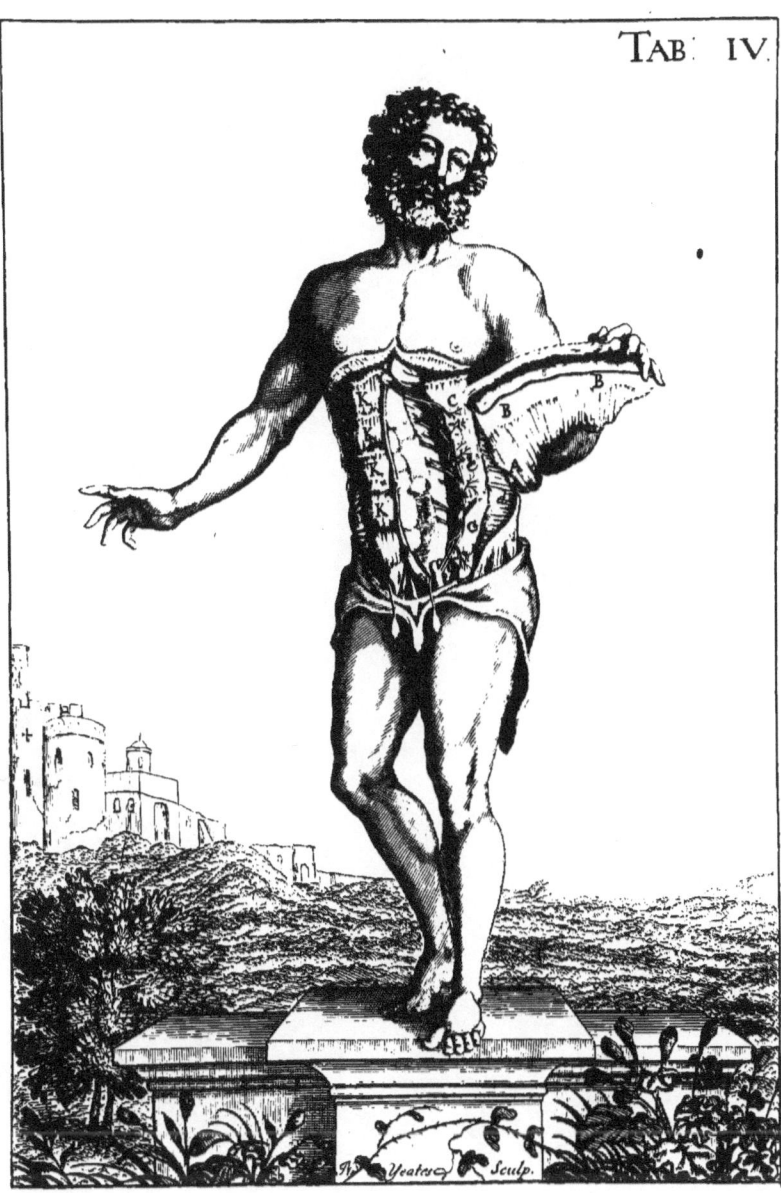

Transversus.

THis fifth Muscle firmly annexed to the adjacent *Perito-* næum is framed of Transverse Fibres, it ariseth from the Transverse processes of the Loyns and the *Os Ileon*, and the Cartilaginous inward part of the lower Ribs with a Nervous Membrane, from the same Ligament as the former; as also with a Fleshy Origination from the inward part of the Spurious Ribs, and Terminates in a broad Tendon at the Semi-lunary Membranes, the *Linea alba*, and so down to the *Os Pubis*; he adheres firmly to the adjacent *Peritonæum* in a manner all along, save only in the *Pubes*, where he divides himself. *Bartholinus* declares, that its chief use is to Compress the Colon. Its generally agreed by all Anatomists, that these Muscles do serve for bringing the Abdomen inwards; the which action is very necessary in the Contraction of the lower Belly. *Spigelius* allows it another action, which is, that it moves the *Thorax* Circularly to the Sides, hereby promoting and cherishing the Native Heat; These Muscles also being of a moderate thickness, do serve as a defence and covering to the subjacent Parts

This brings the Abdomen inwards.

This Transverse Muscle is obliquely perforated a little above the Oblique ascending Muscle, by the Cremasters, and Spermatick Veins and Arteries; so that the perforations of the Oblique Descending, and Ascending Muscles of the Abdomen being framed one above another, not in streight but in bevil Lines, do intercept the passage of the Intestines falling into the Scrotum. These Transverse Muscles running the breadth of the Abdomen cross-ways, do run counter to the Fibres of the Right Muscles, which binds in the Intestines crossing the length of the Abdomen downwards long-ways, as the Fibres of the Transverse Muscles running overthwart, do in their Right Angles secure the Intestines in their actions broad ways, and cross the Abdomen.

Obs.

The Explanation of the Fifth Table.

A. *The Inward face of the Right Muscles.*
B.B. *Another of the same cut in pieces.*
e.e. *The Pyramidal Muscles laid bare.*
F.F. *The whole Muscle,* Tranversalis *in situ.*
e.e. *Shews its first beginning.*
G.G.G. *Nerves sent to this Muscle from the Spinal Marrow.*
A.A.A.A. *Branches of Veins and Arteries of the Eight Muscle, transmitted into this Muscle.*
I.I.I. *The same Muscle laid bare.*
h. *Shews its Tendon laid bare also.*
K.K.K. *Part of the Peritonæum to which part of this Muscle doth closely adhere.*
L. *The Navel.*
O.O O. *The Intercostal Muscles.*

Frontalis.

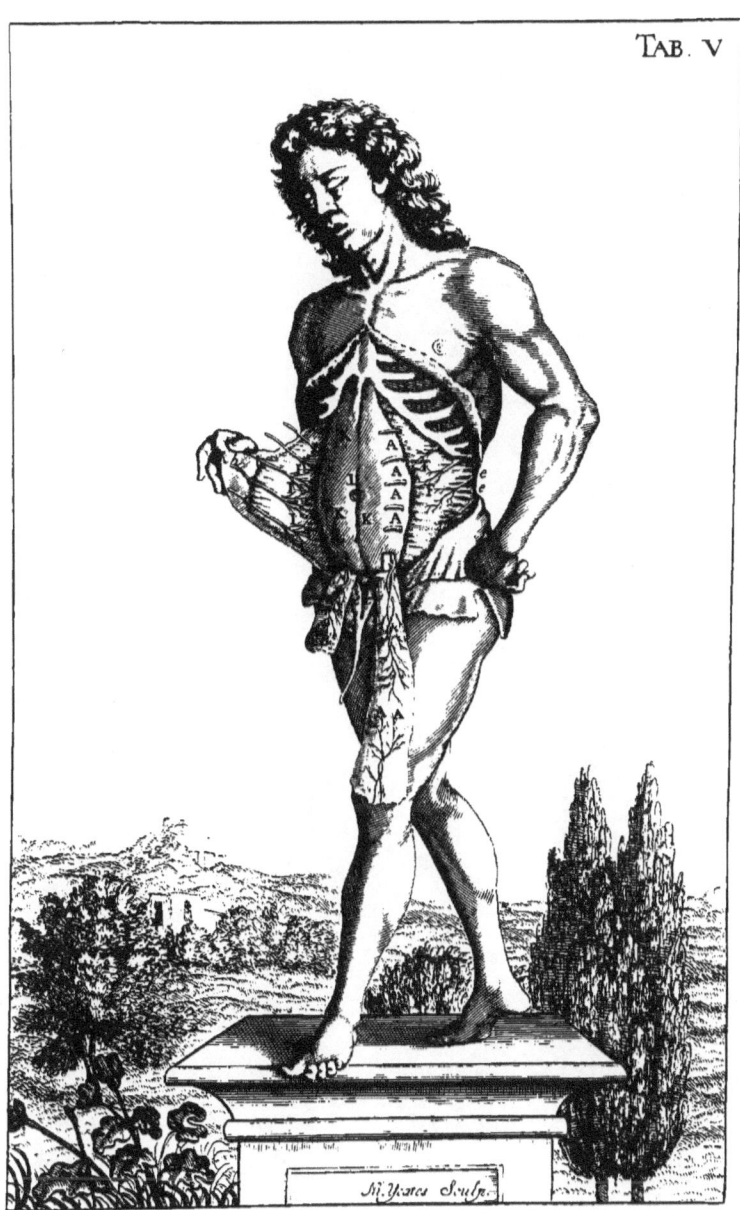

TAB. V

Frontalis.

This Muscle doth lift up the Forehead.

Having cleared all the Abdominal Muscles, which in necessity did require the first use of the Knife: we next come to those parts which in order of Dignity do deserve our Inspection; And here begin we with *Frontalis*, which Muscle doth act variety of postures in Human Bodies, and hereby are excellently delineated the Pictures of Sadness, Joy, Sorrow, and Mirth, these being the Passions of the Mind, and are by this Muscle extreamly well Decyphered. This Muscle ariseth from the most elated part of the Forehead, where the Hair ends, near the Temporal Muscle, and where the Carnous Membrane adheres most firmly to the Cranium, and running right down, is inserted into the Skin which covers the Eyes, and into the Eyebrows; In raising this from the Cranium, you will find Nerves sent from the Cranium to him.

Use and Caution.

This being raised up, it opens the Eye with it: its framed of Right Fibres, and therefore lies as a good Caution for every young Chyrurgeon that he makes no transverse Incision here, lest by neglect thereof, he purchase the utter downfall of the upper Eyelid. To this Muscle by late Anatomists are added the two *Occipitales* which do arise from the middle of the Occiput, *use &c.* and passing by the Muscles of the Ears, tends to the aforesaid *Musculus Frontalis*: though these do not appear so well in all Subjects, yet they do counterpoise in some measure. The proper use of these is to draw the Skin backward, and to keep the Forehead streight and smooth, that so it may not be too subject to Wrinkles, by the frequent action of the former.

This Muscle you have very well described at *Tab.* 6. *Fig.* 1. at *A. A.* at *Tab.* 7. *Fig.* 3.

1. At *A.* Shewing the same *in situ.*
a a. Shewing its beginning.
V. V. Its Termination.
A. At *Fig.* 2. The same Table shews the same Muscle.

Aperiens.

Aperiens Palpebram Rectus.

This lifts up the upper Eyelid.

THere are two Eyelids belonging to either Eye, the upper of which is moved upwards and downwards, both for the opening and shutting of the same; the first designed for Sight, and the second appointed for maintaining its Vigour. This Muscle ariseth from the upper Orbite of the Eye, near its Elevator, where the Optick Nerve hath its transmission: arising with a thin and fleshy beginning, and is expanded with a broad and thin Tendon into the Margent of the Palpebra, and taking the same up, doth open the Eye; this is not to be shewn but with the other Muscles of the Eye after it is taken from the Cranium.

Obs.
This Muscle and the next following, have their distinct Originations and Insertions, although their Muscular Fibres do seem to unite, which hath made some Physicians conclude them to be both but one Muscle; the which on the contrary by acurate Demonstration, is shewn that they have two distinct Nerves taken into two distinct places; for the upper takes a small Nerve from that Motion which arises out through the Cavity of the upper Orbite, the lower assuming its small Nerves from that which it produceth through the Cavity of the lower part of the Orbite, the which Physicians have observed in a Cymical Spasm of the Face: wherein the lower Eyelid is seen to appear as it were immoveable, convel'd and drawn downwards, whilst the upper doth move Naturally; The which could not succeed if they both should be moved by one Muscle.

This Muscle you have very well explained at *Tab.* 7. *Fig.* 1. at *B. C.* and at *Tab.* 6. *Fig.* 2. at *I. I.*

Claudens.

Claudens seu Semicircularis superior Deprimens.

THis ariseth with a sharp beginning from the inward Angle of the Eye, and from part of the Eyebrow near the Nose, and so falls somewhat transverse towards the Exteriour part of the Eye, becoming fleshy, and so broad, as that he possesseth the whole space between the Eyebrow and the Cartilage out of which the Hairs of the Eyebrow do grow, and is inserted into the Angle of the Eye. *This shuts the Eye.*

Julius Casserius, Sect. 1. *de Organ. cap.* 8. doth describe these Indications drawn from the Palpebra, as that such as have their upper Eyelids elated, to be Proud and Fierce; those who have them deprest, to have their Eyes as it were half shut, these generally accounted to be of a mild and humble Nature and Disposition. *Use.*

This Muscle you have described at *Tab.* 6. *Fig.* 2. at *I. I.* at *B. B. Fig.* 1. at the same Table, and at *Tab.* 7. *Fig.* 1. at *B.* and *B. Fig.* 2.

Inferior seu Attollens Semicircularis.

The nether is lifted up by this.

THis ariseth being Membranous and thin from the top of the Nose near the Angle of the lower Eyelid, with a sharp point, and carried transverse about the middle of the Lid, becoming fleshy, and is inserted into the same Angle as the other. Amongst Authors there is held a Dispute, whether that these two Muscles be not properly one, and that *Orbicularis* being so generally received; though the one doth depress, and the other attol or lift up, at the same time to make a perfect close over the Eye.

These Muscles are as Drawbridges for the Eye to open and shut; and as the first doth produce its perfect Sight by being drawn up, so doth the other procure its ready Rest and Quiet by as perfect a closure.

Use.

To these *Riolanus* doth add *Musculus Ciliaris*, which he would have serve to the more firm closure of the Eyelids; but this depends rather on his Fancy or his Knife, than any Foundation in Nature: nor is there need of such a Muscle, since that Office is sufficiently performed by the two forementioned; if they be not one Orbicular Muscle, as some suppose, and consequently by their Circular Fibres constringe as firmly as a Sphincter does.

Both these you have described at *Fig.* 1. *Tab.* 7. at *C.* and at *B. Fig.* 2. in the same Table.

Rectus

Rectus Oculi Primus Elevator.

THe Curtains of the Eye being drawn, we next arrive at the Body of the Eye its self with its Muscles, by which it performeth all its variety of Motions; the truth is, so much hath already been said of these Muscles, as well as of the Coats and Humours which belongeth to the Eye, that its lost time to trouble you with Repetition thereof; this only I shall offer, that this fine Globulous Body ought to have so many Muscles as it enjoys, to make it hang so perpendicularly in the Orbite of its Bone, and be so well ballanc'd, as that it may perform every of its Motions with dexterity, and thereby with ease and advantage execute those Offices for which it was at first designed. *This lifts the Eye upwards*

This Muscle ariseth from the upper part of the Orbite of the Eye, near where the Optick Nerve comes forth, and is inserted into that Coat of the Eye called *Cornea*, where it is clear, and near the Iris by a thin and Membranous Tendon.

This Muscle amongst Anatomists is generally accounted the first of the Eye, and is either called *Attollens* or *Superbus*, being held the Master of Pride and Disdain; the expressing of which Action occasions the Eye to open. Dr. *Willis Fol.* 149. *de Anima Brutorum*, doth attribute this Motion to the Eyelid rather, and by him is called *Musculus Humilis*, because in a devout and intense Prayer it is much lifted up, and hence is it that we commonly see the Hypocritical Tribe, who do affect a form of Sanctity to lift their Eyes so much up, that scarce any part but the White thereof is seen, the Pupil in these Creatures being so obscured that scarce any part thereof is to be seen. *Use.*

This you have very well delineated at *Tab.* 6. *Fig.* 3. at G. G. and at *Fig.* 5. in the same Table at *A*.

Rectus Secundus five Depreffor.

This brings the Eye downwards.

THis is by fome Anatomifts called the fecond Mufcle of the Eye, and is fo far from advancing it, that it brings it down; and hence hath it the name of *Humilis* given it, by fome alfo it is called *Deprimens*, fhewing by its Action and Motion the Index of an humble and fubmiffive Temper and Conftitution; it is oppofite to the former, and is leffer, arifing from the lower oppofite part of the fame Cavity, and hath its infertions as the former, it brings the Eye towards the Chin; This Mufcle you have exactly defcribed in *Tab.* 6. *Fig.* 3. at *H.* and at *D. Fig.* the 5*th.* of the fame Table.

Use.

Obs.

☞. This Mufcle is lefs than the former, which is its Antagonift Mufcle, though their powers hereby are not rendred unequal; becaufe there is lefs force required to deprefs than to elevate, and therefore we fee it happen in like manner in other parts of the Body. And indeed our great Architect has hereby admirably provided for a due balance of the oppofite Mufcles; for where there is an excefs of power or action on either fide, we ufually fee a Spafm fucceed: as happens in the following Mufcle, (or *Adducens*) which in Children is fo often contracted by a vitious turning their Eye inward upon drinking, or otherwife, that it occafions that deformed Squinting or caft of the Eye, the which on the account of the Pliablenes of the oppofite Mufcle they are fo prone to in their tender years, over what they are in a more confirmed Age.

Rectus

Rectus Tertius five Adducens.

THis third Muscle of the Eye doth arise from the Orbite *This brings the Eye inwards.* of the Eye near the Origination of the Elevator, subsisting in the inward Angle, drawing the Eye inwards towards the Nose; This Muscle by some Authors is called *Bibitorius*, and amongst good Fellows great respect is given to this Muscle, bringing their Eye towards their good Liquor; bringing the Eye towards its inward Angle, and making it hereby look somewhat a Squint. Dr. *Willis* in his Book *De Anima Brutor. cap.* 15. *de Visu*, makes mention of a Young man troubled with the Palsie, who when his other Muscles of his left Eye were relaxed, this *Adducens* was strongly contracted, and hence his Eye was so distorted, that its Object seemed as it were double, neither could he distinguish any thing very well with it.

This Muscle you have excellently described at *Tab.* ⁀. *Fig.* 2. at *G.* at *Fig.* 3. at *I I.* in the same Table, and also at *Fig.* 5. of the same Table at *B.*

E Rectus

Rectus Quartus five Abducens.

This brings the Eye outwards.

THis Muscle by Anatomists is sometimes called *Abducens*, and likewise *Indignatorius*, from its cross and scornful Effects it carries with it, bringing the Eye outwards. This Muscle ariseth from the External Angle of the Eye, and is inserted as the former: the Eye is drawn inwards by these four working together, and the motion is as it were suspended, which by Physicians is called a Tonick Motion. In Dissection of a Monkey, all these Muscles perfectly appeared exactly as in a Human Eye: and which is observable in these Creatures distinguishable from all other Bruits, these have no *Musculus Suspensorius*, or *Septimus Brutorum*.

Obs. &.

This Muscle you have also exactly described at *Tab.* 6. *Fig.* 2. at H. and at K. K. *Fig.* 3. and at C. *Fig.* 5. of the same Table.

Obliquus

Obliquus Minor five Inferior.

THis Mufcle arifeth from a Chink which is in the lower part of the Orbite of the Eye, in his Origination Flefhy, fmall, and not altogether round, and is carried Obliquely in his whole courfe, and afcending by degrees to the upper part of the Eye, is there inferted by a fhort but Nervous Tendon, near the Tendon of the *Abducent* Mufcle; not far off which, the Tendon of the other Oblique Mufcle hath alfo its infertion, and by moving the Eye downwards, it doth convert and abduce it towards its External Angle in a rowling manner.

This brings the Eye towards its External Angle.

Ufe.

This Mufcle alfo you have defcribed at *Tab.* 6. *Fig.* 3. at *L.* and at *Fig.* 5. at *E.* in the fame Table.

Obliquus Major vel Superior cum Trochlea.

This carries the Eye to its inward Angle.

THis ariseth from the same place with the *Abducent*, and marcheth in a right Line to the External part of the Internal Angle, where it grows indifferently thick, and is then attenuated and grows round, and goes through a Pulley there designed for it, the which so soon as it hath past it, it, yet so bends its self that it makes the Right Angle of the Eye; and running upwards it begins to grow Oblique, and passing by the *Levator* is inserted between the *Abducent* and Oblique Muscles, as is before demonstrated. The *Trochlea* or Pully is a perforated Cartilage, passing to the Bone of the upper Mandible, near the inward Angle of the Eye; These two Muscles are called *Amatorii*, or the Lovers Muscles, being as the true Messengers of Affection, by some they are called *Circumactores* or the Rowling Muscles, for they do much work in Human Body; in Sheep also they are of very great use, being given them as their chief Watchers, or those which do work their Eyes about. The young Chyrurgeon is here cautioned, that in his Curing of

Caution. *Fistula Lachrimalis*, he takes great care of this Muscle. The
Use. use of this Muscle is to turn or rowl the Eye inwards towards the inward Angle of the Eye by a Circular Motion, and so to
Obs. &. abduce the Pupil from the *Nares* or Nostrils; several Fibres do pass from the *Periostium*, or inward Film of the Orbite to the forementioned *Trochlea*, and according to some do constitute the *Musculus Trochleæ*, though the use of them seem designed for a steddy fixing the *Trochlea*, rather than for a Muscular motion.

This Muscle you have also exactly described at *Tab.* 6. *Fig.* 2. at *D. E. F. D.* shewing the Muscle *E.* the *Trochlea*, and *F.* the chord of this Muscle; at *Fig.* 3. at *M. M.* you have it, and at *E. Fig* 5. you have it again very exactly delineated.

Attollens

Attollens Aurem.

AS the Eyes are placed in the forepart of the Head as *This carries* Watchmen to guard the Body, so also hath our wise *the Ear up-* Creator planted two Ears at the sides thereof for the *forwards.* better perception of Sounds, and a more ready passage to hearing, to the better performance of which there are hereto given variety of Muscles; amongst which, this is reckoned as the first, it arising from the External Termination of the Frontal Muscle, and so being thin and Membranous, is carried over the Temporal Muscle, and is inserted growing narrower into the upper part of the Ear, moving it upwards and forwards. *Use.*

This Muscle you have exactly delineated at *Tab.* 6. *Fig.* 4. at H. and at S. *Tab.* 7. *Fig.* 1. you have the same again.

F Detrahens

Detrahens Aurem.

This moves the Ear downwards.

THis Muſcle ariſeth Fleſhy, broad, and ſometimes Fibrous, from the back part of the Head, near the Mamillary proceſs, and ſo growing narrower in its progreſs is inſerted into the whole Cartilage which encompaſſeth the Ear; be careful in raiſing the *Cutis* left you take up this Muſcle with it, and ſo looſe him; this moves the Ear downwards and forwards, this Muſcle is by ſome allowed as part of *Quadratus Buccas Detrahens.*

Uſe.

This you have ſhewn you at *Tab.* 7. *Fig.* 1. at *T. T. T.*

Adducens

Adducens Aurem ad Anteriora.

THis is a common Muscle, being part of that which *Spi-* *This draws*
gelius calls <u>*Quadratus Buccas Detrahens*</u>, and is also al- *th. Ear*
lowed as part of that Muscle called *Platusma Myodes*, *forwards.*
from whose insertion you will find a Fleshy and Fibrous Elongation implanted into the Root of the Ear.

    This Muscle is said to draw the Ear forwards and somewhat *Use.*
upwards. Expect to meet this at <u>*Quadratus Buccas Detrahens*</u>,
as being a part of it.

Abducens

Abducens Aurem ad Posteriora.

This brings the Ear backwards.

THis Muscle is planted at the *Occiput*, and ariseth above the Mamillary processes from a Knot of Muscles which belong to the *Occiput*, with a narrow beginning, and being carried downwards transversely, is inserted with a double and sometimes treble Tendon into the hinder part of the Ear. This Muscle is said to draw the Ear backwards: In Beasts, as in Horses, Oxen, and the like Bruits these Muscles are much more large and apparent, and oft times more numerous, whence it is, that these can move their Ears more powerfully, and act those strong motions with them which we see is customary for them to do.

This you have described at *Tab. 6. Fig. 4.* at *I. I.*

Externus

Externus Tympani Auris.

THe inward Ear hath two Muscles allowed it, found out *This brings the Tympanum upwards and forwards.* by the two excellent Anatomists *Hieronimus Fabritius de aqua pendente, & Julius Casserius Placentinus,* and of these one is planted outwards, the other inwards; and from hence they do take their names; this ariseth from the upper and inward passage of the Auditory passage with a large Origination, and becoming Fleshy is inserted externally by a short Tendon into the *Tympanum,* extending the Membrane with the *Malleus* upwards and outwards. *Spigelius* saith it is small and *Use.* ariseth from the *Cutis,* and that Membrane which covers the Auditory passage.

This Muscle is one of the least in the whole Body, and is &c. to be shown entire with some difficulty; great caution therefore is to be used in opening the *Os Petrosum,* about that part which respects the Temples, whether it be done by the small Chissil or Filing, that so the pieces of Bones being taken out by degrees, this Muscle may receive no prejudice; the like care is to be observed also in shewing the following Muscle.

This Muscle you have exactly shewn you at *Tab.* 6. *Fig.* 6. at *A*.

G   Internus

Internus Tympani Auris.

This brings it obliquely forwards and somewhat inwards.

THis Muscle is inwardly planted, seated in the *Os Petrosum*, having its Origination from the Basis of the *Os Cuneiforme*, and so becoming Fleshy, though thin and small, and having made half his progress, divides himself into two very small and very thin Tendons, the one of which is implanted into the upper process of the *Malleus*, the other into the Neck thereof, drawing it obliquely forwards, and bringing it somewhat inwards.

Use.

These two Muscles do then first move the Membrane with its small Bones upwards and downwards, when we would carefully listen or hearken to any important Matter, *&c.* Matter or Concern, as *Diemerbroeck* observes. Dr. *Willis Fol.* 133. de

Obs.

Anima Brutorum, writes, that the action of this Muscle is involuntary, and is wrought about by some Instinct of Nature; for when a very vehement sound doth approach the Ears, this Muscle doth remit its great noise, so as that it does sensibly obtund the relaxed *Tympanum* more powerfully; but if it be either more thin, or more obtuse, this contracted Muscle doth distend the *Tympanum*, so that this obtuse Impression may be made more sensible.

This Muscle also you have exactly described at *Tab.* 6. *Fig.* 6. at *C. C.*

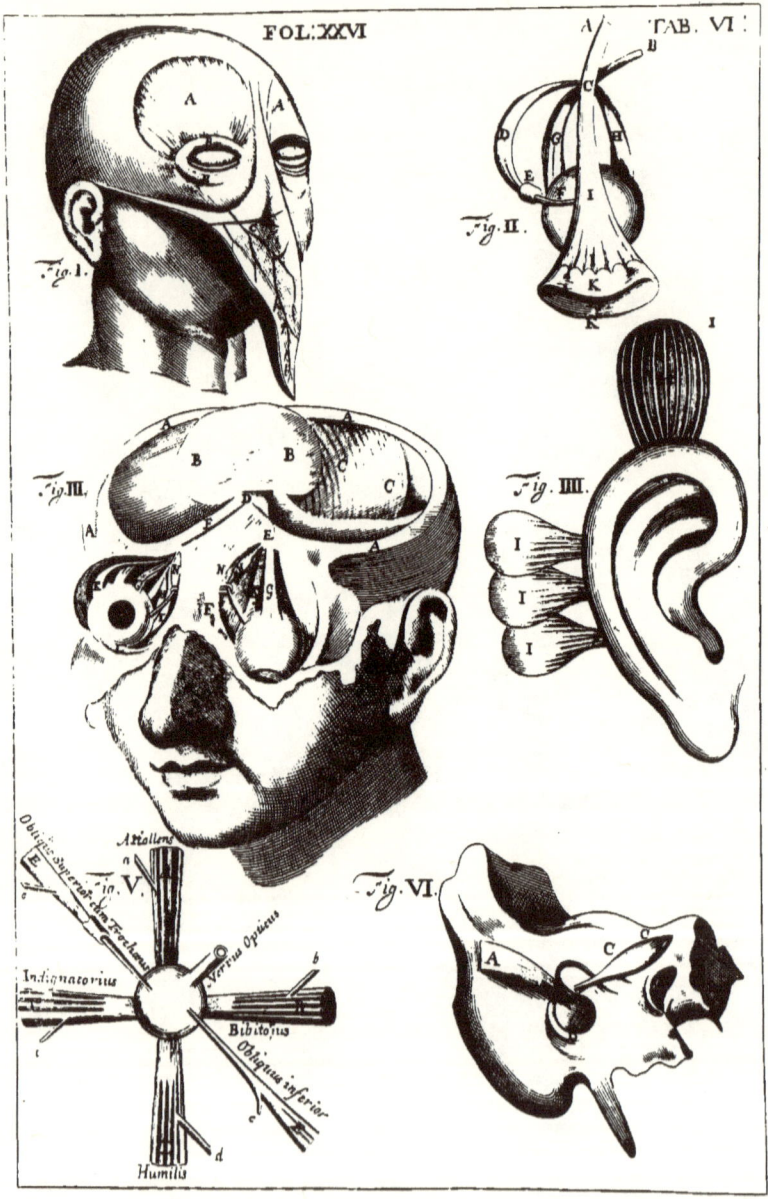

The Explanation of the Sixth Table.

FIG. I.

Shews at *A. A. The Frontal Muscle.*
B. B. *The Muscles of the Eyelids.*
b. c. *The* Membrana Carnosa *laid bare.*

FIG. II.

A. *The Optick Nerve.*
B. *The Motory Nerve.*
C. *The Exortion of the Muscles.*
D. Musculus Trochlearis.
E. *The* Trochlea.
F. *The Chord of this Muscle.*
G. Musculus Adducens.
H. Musculus Abducens.
I. I. *The Muscle of the upper Eyelid.*
K. K. *Shews them cut off.*

FIG. III.

A. A. A. A. *Shews the* Cranium *resected.*
B. B. *Part of the Brain.*
C. C. *The* Cerebellum.
D. *The uniting of the Optick Nerves.*
E. E. *Their progress to the Eyes.*
G. G. Musculus Attollens.
H. Deprimens *of the Right Eye.*
I. I. Adducentes.
K. K. Abducentes.
L. Obliquus Externus.
M. M. Obliquus Internus.

FIG. IV.

H. Musculus Auriculam recta sursum Movens.
I.I.I. Musculus Triceps eandem sursum trahens.

FIG. V.

A. Attollens.
a. *Shewing its Nerve.*
B. Bibitorius.
b. *Shews its Nerve.*
C. Indignatorius.
D. Humilis.
E. Musculus Trochlearis, *or* Obliquus Superior.
F. Obliquus Inferior.
o. *The Optick Nerve.*

FIG. VI.

A. Externus Tympani Auris.
B. Membrana Tympani.
C. C. Internus Tympani Auris.

Abducens

Abducens Nasi Alas.

This dilates the Nose.

THe Nose is the Organ of Smelling, and this gains place in the more eminent part of the Body, for the better susception of the ascent of Vapours and invisible Halations, and their Qualities are sent up hither by the Olfactory Nerves to the common Sensory, and are there approved of according to Judgment; and thus is Man brought into a capacity of either taking or refusing such things as may prove either beneficial or prejudicial, purchased without the Eye, and perceived without the Ear; The Nose is not wholly moved, but rather its lower parts which we call *Alæ*, and these are either kept open or shut by the benefit of Muscles: and these are called either *Abducentes* or *Adducentes*, or if you please *Aperientes* or *Claudentes*.

Use.

This Muscle ariseth very small and Fleshy from *Os Maxillæ Superioris*, near the first *Par Labiorum proprium*, and is inserted into the lower and upper part of the *Alæ*, and moveth either part upwards.

This you have exactly delineated at *Tab.* 7. *Fig.* 1. at *D. in situ*, *e.* shews its Origination, *d.* its Insertion *Fig.* 2. *id. Tab.* you have the same at *B.*

Attollens

Attollens Nasi Alas.

THis Muscle ariseth from the top of the Bone of the Nose, near the *Lachrymal* Cavity, with a sharp and Fleshy beginning, descending to its sides in a triangular form much resembling the *Greek* Letter Δ. and so descending according to the length of that Bone, is inserted broad and Fleshy into the *Alas Nasi*, and do carry the same upwards; *Casserius* hath found them much resembling Myrtle Leaves, these two pair drawing the *Nasi Alas*, do dilate and open the Nostrils. *This brings it upwards.*

Use.

This you have fairly described at *Tab.* 7. *Fig.* 1. at *E. e.* shewing its sharp and Fleshy beginning, *f* shewing its Termination into the *Alas*, B. shews the same in the same Table *Fig.* 2.

H   Claudens

Claudens Nasum Externus.

This Muscle shuts the Nostrils.

THese Muscles are very small, and without it happen in very Nasute Persons they are never seen or to be distinguished; the first pair of which is outwards and Fleshy, and so do arise at the Root of the *Alæ*, and so climbing transversely is inserted into the tip of the Nose; and as *Veslingius* judgeth, doth dilate the same, whilst others do affirm that this doth constringe it.

Use.

These Muscles are not to be shewn by any Figure, being both so small, and so inwardly implanted.

Claudens Nasum Internus.

This is much like the former as to its bigness, lodging in- *This doth* wardly under the Membrane which covereth the Nose *constringe the Nose* from the Bone, it ariseth from the end of the Bone of *inwards.* the Nose, and is expanded into the *Alam Nasi*, and doth constringe it; this is very small, and very rarely found out, unless in such Nasute Persons whose general Series of Muscles are very apparent, thicker, and larger than ordinary.

There is also another Constrictive Muscle, which hath gotten the name of *Orbicularis* common to the upper lip, the which drawing the Lip downwards, doth also therewith shut up and close the Nostrils; And this *Bartholinus* describes in *Fol.* 358. *Anatom.* where he affirms, That he hath observed an Appendix hereof to descend to the upper Lip, and that in such People who could not lift up their Nose without their Lips.

This Muscle with its former is not to be shewn by Figures, they being so very small.

Zygomaticus

Zygomaticus Riolani, vel Attollens Labium Superius.

This lifts up the Lip.

TO Man are given Lips, both for his accommodation of Eating and Drinking; as also for forming his Voice, retention of Spettle, shutting the Mouth and defending the same from outward Injuries; and because all these Qualifications do require a voluntary Motion, they have given to them variety of Muscles to perform the same: amongst which this is nominated one of the third proper pair called by *Riolan. Zygomaticus.*

It ariseth Fleshy from the *Os Sygoma*, with a Fleshy and broad beginning, and running obliquely downwards and forwards is inserted into the side of the upper Lip near *Primus Nasi*, the which doth abduce it to its sides upwards. See this *Riolanus Anatomy*.

This also you have shewn you at the Letter *G.* at *Tab.* 7. *Fig.* 1. and at *Fig.* 2. *id. Tab.* at *C.* you have the same.

Abducens

Abducens Labia.

THis ariseth from the *Os Zygoma* Fleshy and round with much Fat, and is implanted into the Lips where they are joyned together: this moves the Lips upwards and outwards, and doth also help the former in their motion; this by some Anatomists is accounted the first proper pair: these Muscles do also draw the Lips to the side, whether either one or both do operate either singly or together. *This brings the Lip upwards and outwards. Use.*

This Muscle you have shewn *in situ*, at *Tab.* 7. *Fig.* 1. at *H. i.* and *k.* shewing both its Origination and Insertion; the same you have at the Letter *D.* in the second Figure of the same Table.

I   Labium

Labium inferius Deprimens.

This brings the lower Lip downwards and outwards.

THis ariseth Fleshy from the lowermost and outermost part of the lower Mandible, whence running obliquely, it is broadly inserted into the middle of the Lip, and moves it downwards and outwards; this is called the fourth pair of the proper Muscles.

This Muscle hath a part with the first pair called *Detrahens Quadratus*, by which it is obliquely moved to either side, (*vid*) right or left, as either of them are in Operation, drawing the whole lower Lip downwards.

Use.   This Muscle you have shewn at the Letter *I. Tab.* 7. *Fig.* 1. and at *E.* in the second Figure of the same Table.

Par

Par Labium conftringens five Mufculus Orbicularis & Conftrictor.

This is common to either Lip, being framed of a fun- *The Lips*
gous Subftance, with Orbicular or Conftrictory Fibres, *are purfed up by this.*
arifing from the middle of the Bones of the upper
and lower Mandible, and doth form and make the whole
Body of the Lips, encompaffing the Mouth like a Sphincter,
and drawing the Lips mutually to it, the which do firmly adhere to the red *Cutis*, which gives them their admirable Vermilion Dye, as alfo fhews the Palenefs of them in Sicknefs; All
thefe Mufcles of the Lips are fo clofely conjoyned to the *Cutis*,
that their Fibres do interfect one amongft another; and hence
is it that fuch a variety of motions are feen in the Lips; and
thus have we fhewn all the Mufcles of the Lips. *Fallopius* hath
another pair of thefe Mufcles belonging to the Lips, whofe
Cutis is fo clofely mixed with the Mufcles, that it rather feems
to be a Mufculous *Cutis* or a Cuticular Mufcle.

This Mufcle is alfo called *Ofculatorius* from the ufe which is   *Ufe.*
made of it.

Thefe Mufcles of the Lips have variety of Ufes given them
for performing their diverfity of Actions; as fome being defigned for fhutting the Mouth, others for opening the fame;
fome made for accommodation of Eating and Drinking, and
others formed for Ornament of Speech and Love Motions.

This you have fairly fhewn you at *Tab.* 7. *Fig.* 1. at K. K.
and at *Tab. id. Fig.* 2. you have the fame at F. F.

Quadratus

Quadratus five Platyfma Myodes.

This draws down the Cheek.

THe firft Mufcle lying under the Skin of the Neck is called *Quadratus*, from its Figure, and is fmall and Membranous firmly adhering to the *Cutis*, arifing from the Vertebres of the Neck, *Scapula Clavicle*, and *Sternon*, large, broad and thin, with ftore of Membranous Fibres, and fo running up with oblique Fibres, is inferted into the Chin, where both the upper and lower lips are joyned; and enlargeth himfelf fo far as to make *Adductor auris ad Anteriora*, and becaufe it agrees much with the Chin, it helps forwards the opening of the Mouth; and by its enlarging its felf fo far as to make *Adductor auris ad Anteriora*, the Ears alfo may be allowed to be moved by the help of this Mufcle; it hath various Surcules of fmall Nerves belonging to it from thofe of the Neck; This Mufcle being once convelled there follows a Cynick Spafm.

Caution.
Here's Caution alfo for the young Chyrurgeon, where he ought to take notice of the Fibres of this Mufcle, efpecially when he may or fhall be called to make Incifion here; for want of knowing of which upon making tranfverfe Incifions here, and croffing the Fibres and their Ductures, he prefently occafions an Avulfion in the Cheeks, otherwife great care muft be had to preferve this Mufcle; for whether you do raife the *Cutis* from above or below, he adheres very clofely to it; In the raifing this Mufcle, be careful of leaving its Elongation that makes *Adductor auris ad Anteriora*, which you will rarely mifs.

This Mufcle according to *Riolanus* you have expreffed in our 10th. *Tab Fig.* 1. at F.

Buccinator.

Buccinator.

THis second Muscle which formeth the Cheek is called *Buccinator* from the use that is made of it; it ariseth from the upper part of the upper Mandible, and from the lower part of the lower Mandible, where the Gums begin, and do amplect the whole Cheeks in their seats, being round like a Circle: the proper Coat of the Mouth adheres so firmly to him inwards, that he is scarce separable from it; outwardly he hath a large round Tendon implanted into the midst of him, which hath his Origination from a Glandulous Substance, growing to the *Os Zygoma*, close by the Origination of the Muscle so called. *This draws the Cheek inwards.*

This Muscle doth not only move the Cheeks with the Lips, but doth also constringe them, and drive the Meat fallen into the Mouth into the Cheeks back again to the Teeth, sending or conveying the Meat thither, until it be better lessened, and made smaller, and a more accurate Confraction of the same be hereby made. *Placentinus* writes, That he hath found a very strong Ligament in the Center of this Muscle; the which arising outwards, and creeping along the *Os Gingivæ*, terminateth into a small and thin Muscle directly opposite to the Cheek; but *Riolanus* denies the same. *Use.*

This Muscle if you please may also be allowed as a hand to the Teeth, sending the Meat till it be well Chewed to the Teeth, by which it may the better be lessened and comminuted. *Use.*

This Muscle you have shewn you at *Tab* 7. *Fig.* 1. at *F.* and at *P. Tab. ead. Fig.* 2.

K   Masseter

Maſſeter ſive Lateralis, ſeu Manſorius.

This draws the nether Mandible laterally.

THis ariſeth with a double beginning, ſtrong, large, and Nervous; firſt from that Suture where the fourth and firſt Bone of the upper Mandible joyneth; ſecondly Fleſhy from the *Os Jugale*, and is firmly and largely inſerted into the lower Mandible External; This Muſcle by reaſon of its diverſity of Fibres given it, doth move the nether Mandible forwards, backwards, and laterally, and as it were about alſo; If you throw this Muſcle either from its Origination or Inſertion, *Temporalis* will appear in its Inſertion.

Uſe.

Uſe. The proper Uſe of this Muſcle is ſhewn in Maſtication; it moves to both ſides, as to Right and Left; it takes its name of *Manſorius* from its proper Action: and from its *ſite*, it is called *Lateralis*.

This you have ſhewn you at *Tab.* 7. *Fig* 1. at the l etter *O*. *I.* ſhews it alſo at *Fig.* 2. *Tab. ead.* where *b. b.* ſhews its Origination, and *c. c.* its Inſertion.

Temporalis

Temporalis seu Crotaphites.

Tis drawn the Mandible upwards.

This is the first and strongest of all the Muscles, filling the whole Cavity of the Temple Bones; it ariseth from the *Os Frontis, Syncipitis, & Sphænoides*, fleshy and Semicircular, and growing narrower in his descention, runneth under *Os Jugale*, with a short but very strong and fleshy Tendon into the process of the lower Mandible called *Corone*; it hath allowed it three Nerves on either side: One from the third, a second from the fourth, and a third from the fifth pair; Wherefore this Muscle being either inflamed, contused or wounded, sharp pains do immediately succeed, and great danger of Convulsion and fear of Death, especially if the hurt do happen about the Nervous part thereof; As touching the *Periostium*, you will find that if you do raise this Muscle carefully, (contrary to the opinion of some) you will meet it under this Muscle, although many have asserted the contrary. The end of this Muscle is the beginning of the lower Mandible, the which it moves and draws upwards, and hereby shutteth the mouth; and by some hence it is called *Attollens*, and is the strongest Muscle of the Body in respect of its bulk. And as *Spigelius* writes, it is furnished with so much strength, that he remembers in his ripe years that he hath lifted up several pounds of Lead by the strength of his Teeth, and carried them therewith. I have seen very great weights brought from the ground by the Teeth and strength of this Muscle, and it has been reported above 200 *l*. weight has been lifted from the ground by them. This Muscle both shuts the Mouth, and constringes the Teeth.

Us.

In opening the Temporal Artery, caution is to be used by the young Chirurgeon, lest he injureth this Muscle, by cutting too deep, whence fatal Convulsions (or at least a prejudice to Mastication) do succeed. The safest way therefore is, first to make a light incision of the Skin, and then the Artery lying bare may the readier be hit and divided without injuring this Muscle, which lies under it; or at least it will be well to use in this Operation the Lancet that is retuse on one side like the Penknife, and to take the Artery pretty high about that part of the Temples which joyns to the *Os Frontis*.

Obs. Cr.

This you have at *Tab.* 7. *Fig.* 1. at *L m. m. m.* shewing its Semicircular beginning, H. shews the same, *Fig.* 2. *ejusd. Tab.il.*

Mastoideus.

Maſtoideus.

This contracts the Neck.

THis is one of the eight pair ſeated in the forepart of the Neck, ſtrong, long, and ſmooth : it ariſeth with a double beginning ; a certain Cavity diſtinguiſhing them, one from the *Sternon* Nervous, and the other Fleſhy from the more elated part of the *Clavicle*, and is obliquely inſerted into the Mammillary proceſs by a round and Fleſhy Tendon.

Obſ. This is worthy Note, that in a live Man, eſpecially in thin and aged People, in which it is very conſpicuous, that if the *Uſe.* whole pair do work, it bends the Head right, but if one of them do only work, this Muſcle is only contracted, and this brings the Head forward laterally.

This you have at *Tab.* 7. *Fig.* 2. at *K in ſitu*, *d. d* ſhewing its Origination from the *Sternon,* *e. e.* its Inſertion into the Mammillary proceſs of the Temple Bone.

Biventer

Biventer Digastricus seu Graphyoides.

This is the second pair called *Biventer*, the like of which is not to be found in Human Body; in the middle of whose Venter it groweth thinner, having as it were two Venters, arising near the Mammiform procefs: it arifeth with a broad and Nervous beginning from the procefs *Styloides*, and suddenly becoming round, Fleshy, and small, so soon as he recurvates under the *Styloides*, he becomes a Nervous round Tendon, and then becomes Fleshy again, and is inferted into the middle Interiour part of the nether Mandible, and by drawing the Mandible downwards doth open the Mouth; the too great descent of which is hindred by its annexed Ligament. And for the better performance of this work, the former Muscle doth help this very much in its Operation. *This draws the Mandible downwards and opens the Mouth.*

Provident Nature hath given this Muscle two Bellies, and from thence it got its name, having in its middle an interjacent Tendon, very aptly refembling a *Trochlea* or Pulley; formed for this use, that whereas these Muscles do not arise from the lower parts of the Neck, but rather bred from the upper parts thereof; they do reflect about the lower parts of the Mandible like a Pulley, drawing the Mandible downwards, and so opening the Mouth. *Ufe.*

This you have shewn at *Tab.* 7. *Fig.* 1. at *N. Extra Situm*, whose Tendon is described by *N.* and its Venters by *L. M.*

L   Coracohyoides.

Coracohyoides.

This brings the Os Hyoides obliquely downwards.

This Muscle is very thin and lon , so that by Anatomists it is held there is not its like to be found in Human Body, for thinness and length ; it is endowed with a double Belly by a distinct Tendon : it ariseth with a Fleshy beginning near the Neck of the *Os Scapulæ* from its process *Coracoides*, and running under the *Levator Patientiæ* of the *Scapula*, is obliquely carried under the *Mastoides*, and there becometh a small round Tendon, and then Fleshy again : and so is implanted into the Horns of the *Os Hyoides*, and draws it obliquely downwards ; If you leave this Muscle in its Origination at the Dissection of the *Levator*, you will find his Origination perfect.

Use.

Obs.

This Muscle hath a double Venter, as its former Companion, that the *Jugular* might not be too much compress by it.

This you have described very exactly at *Tab. 7. Fig. 2.* at the Letters *L. L.* these two Letters demonstrating its two Venters described in their Natural Position ; and at *Tab. 8. Fig. 3.* you have the same laid bare at *H. H.*

The

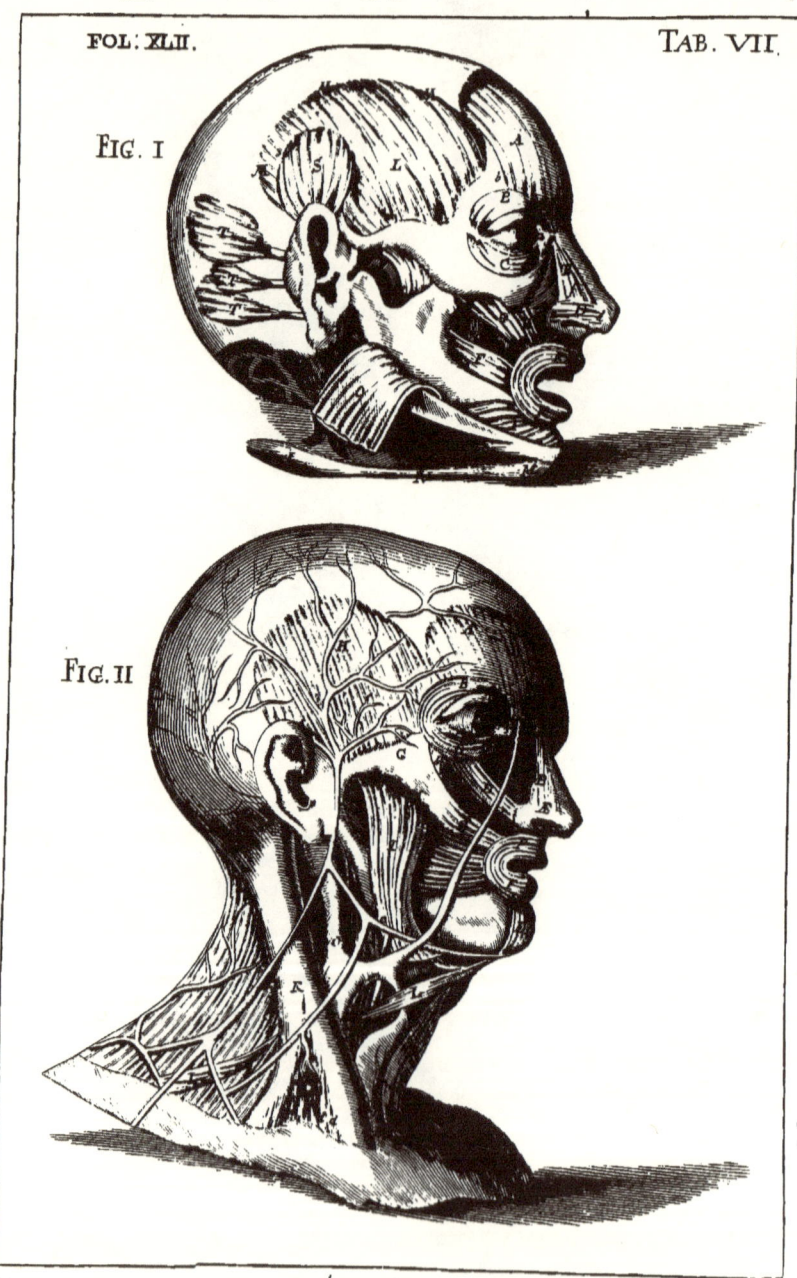

The Explanation of the Seventh Table.

FIG. I.

A. *The Frontal Muscle in situ.*
   a.a. Shews its beginning.
b.b. Its end or Termination.
B.C. The Muscle of the Palpebræ.
D. Abducens alas in situ.
c. Shews its Origination.
d. Its Termination.
E. Attollens Nasi Alas.
e.f. Shews its beginning and ending.
F. Buccinator.
G. Attollens Labium Superius.
i. k. Shews its beginning and ending.
H. Shews the second pair, Abducens Labia.
i. k. Its beginning and ending.
I. Labium inferius Deprimens.
K. K. Labium Constringens.
L. Temporalis.
m. m. m. Shews its Semicircular beginning.
O. Lateralis.
S. Attollens Aurem in situ.
T.T.T. Detrahens Aurem.

·FIG II.

A. Musculus Frontalis in situ.
B. B. The Muscles of the Palpebræ.
Æ. Nasi Alas Constringens in situ.
B Nasi Alas Abducens.
C. Attollens Labium Superius.
D. Abducens Labia.
E. Inferius Labium Deprimens.
F. F. Par Labia Constringens.
H. Temporalis.
b. b. Shews its two beginnings.
c c. Shews its insertion into the largeness of the lower Mandible.
K. Mastoides.
d. d. Shews its beginning from the Sternon.
e. d. Its insertion into the Mammillary process.
L. L. Coracoides.
O. The outward Jugular Vein.
P. Buccinator.
G. Shews the Os Jugale.

Sternohyoides.

Sternohyoides.

This brings the Os Hyoides downwards.

These are generally accounted the second pair, moving the *Os Hyoides* downwards and backwards; This Muscle ariseth broad and Fleshy from the inner part of the *Sternon* under the Skin of the Neck, and running on (the same in substance and breadth all along) the *Aspera Arteria*, and the *Thyreoidal* Cartilage of the *Larynx*, is inserted into the Basis of the *Os Hyoides*.

Use: This Muscle brings the *Os Hyoides* directly downwards and somewhat backwards.

You have this fairly delineated at *Tab.* 8. *Fig.* 2. at *C. C.* and at *G. G. Fig.* 3. *id. Tab.*

Sternothyreoides.

Sternothyreoides.

These Muscles of the *Larynx* (as they call the Head of *the Afperia Arteria*) are so nominated, not because they do move the whole *Larynx*, but its Cartilages; This formerly was called *Bronchus*, but we more properly from its name and insertion do call it *Sternothyreoides* : it ariseth Fleshy and broad from the upper and inner part of the *Sternon*, and keeping his dimensions, creeps up with streight Fibres along by the *Aspera Arteria*, and is inserted into the lower side of the Scutiformal Cartilage, the which when it presses the *Scutiformis*, it narrows its *Rima* or Cleft.

It is generally believed that this does close up the lower part of the *Scutiformis*, and that it draws downwards, whence the upper part thereof is seen to be extended, and the *Rimula* or Cleft dilated.

This also have you delineated at *Tab.* 8. *Fig.* 2. at *L. L. N. N.* shews the same at *Fig.* 3. *Tab. ejusd.*

M   Hyothyreoides.

(46)

Hyothyreoides.

This contracts the Larynx.

THis ariseth from the whole side of the *Os Hyoides*, at his Basis, and running down broad with right Fibres is inserted into the lowest and lateral part of the *Scutiformal* Cartilage, and by attolling it, doth dilate its *Rima*. *Riolanus* thought this pair not to be peculiarly appointed to any Cartilage, but did lift up the whole *Larynx*; if you raise this Muscle clear from its Origination and Insertion, you will be less troubled to find out these Muscles of the *Larynx*.

Obs. This draws the *Larynx* upwards, whence it is we in our selves, when we would frame a sharp Voice, that we then do bring the *Larynx* upwards, and when this is contracted, it contracts the upper part of the *Scutiformis*, driving it inwards.

This also you have delineated at *Tab.* 8. *Fig.* 2. at *m. m. M. M.* shewing the same at *Fig.* 3. *ejusd. Tab.*

Styloceratohyoides.

Styloceratohyoides.

THis third pair ariseth from the Root of the *Processus* *This brings the Os Hyoides obliquely upwards.* *Styloides*, and being small and round, is implanted into the Horn of the *Os Hyoides*, found infallibly by *Digastricus* his running through or under him, obliquely, moving the *Os Hyoides* obliquely upwards.

This draws it obliquely upwards. *Use.*

Observe that its Insertion is in the lower part of the Horn (or rather towards the Basis) of the *Os Hyoides*.

E. E. Shews this, *Tab.* 8. *Fig.* 3. F. F. Shews the same, *ead. Tab. Fig.* 2.

Amongst these Muscles of the *Fauces*, the two pair *S.* lately found out by the Ingenious Doctor *Croune* may not improperly here be mentioned; one of which are named by him *Musculi Pterygo-palatini*, and the other *Spheno-palatini*. The former of these are seated in the Interior part of the Cavity of the *Os Pterygoides*, and terminate with their Tendons (which run on part of the foremention'd Bone as on a *Trochlea*) about the *Glandula palati*, which (together with the *Uvula*) they depress.

The latter of these, or *Spheno-palatini*, arise from the *Os Sphenoides*, and with a broader Tendon are inserted into the sides of the *Glandula palati*, which (together with the *Uvula*) they do *attoll*. From the situation and action of these latter Muscles may some account be given, how the *Uvula being relaxt* is commonly reduc't by thrusting the Thumb bent toward the Palate or these Muscles.

These Muscles you have exactly shewn you at *Tab.* 10. *Fig.* 2. at O. O. X. X. Shews its Tendon, Q. Shews the latter, where *f. f.* shews its Tendon also.

Milohyoides

Mylohyoides Riolani.

This moves the Os Hyoides directly upwards.

THis ariseth laterally from the nether Mandible inwards, under the *Dentes Molares*, Fleshy, and is inserted into the Basis of the *Os Hyoides*, externally; this is to be thrown upwards in Dissection. Look into *Riolan.* which doth give you satisfaction as to this Muscle.

Geniohyoides.

Geniohyoides.

THis first pair which from their primary use were called *Rectà Atttollens*, & *Geniohyoides*, drawing it directly upwards and somewhat forwards, it ariseth internally from that middle part of the lower Mandible called the Chin, and marching down short, broad, and Fleshy, is inserted in a proper Cavity, at the Basis of the *Os Hyoides* internally. *This draws it upwards and forwards*

This Muscle moves the *Os Hyoides* directly upwards, and somewhat forwards. *Use.*

D. D. Shews this Muscle, *Tab.* 8. *Fig.* 3. E. E. Shews the same, *Tab. ead. Fig.* 2.

N   Mylogloſſus.

Myloglossus.

This moves the Tongue upwards.

THe Tongue, whereas it is not only the primary Inſtrument of the Voice, but alſo is uſeful for turning of the Meat contained in the Mouth, and doth obtain very many voluntary Motions; for the executing of each of which, there are required ſeveral Muſcles, amongſt which this is ſaid to be the fourth pair, it ariſeth with a broad beginning from the innermoſt lateral part of the lower Mandible under the *Molares*, and is inſerted into the Ligament which ties the Baſis of the Tongue to the *Fauces*; At the Origination of *Mylohyoides* you will certainly find this, and it is beſt ſhewn when the Mandible is divided: when one of theſe move, the Tongue is turned upwards: when both move, the tip is directly lifted upwards towards the Palate.

C. Shews this Muſcle, *Tab.* 8. *Fig.* 2.

Ceratoglossus.

Ceratoglossus.

This is one of the four pair arising from the Horns of the Os *Hyoides*, and hence it is called *Ceratoglossus*, and is implanted obliquely into the sides of the Tongue, near its Root; if both these work together, they draw the Tongue downwards and inwards: if only one operate, it moves it either to the right or left side. *This brings the Tongue downwards. Use.*

I. Shews this, *Tab.* 8. *Fig.* 3. *D.* Shews the same, *Fig.* 3. *ejusd. Tab.*

Genioglossus.

Genioglossus.

This moves it forward.

Vf.

THis is one of the second pair so called by *Spigelius* : it ariseth with a narrow Origination, about the middle of the lower Mandible or Chin; and then enlarging himself, is inserted into the Root of the Tongue ; *Veslingius* doth number this amongst the number of *Os Hyoides*, and saith that they are implanted at the Basis of the Bone, which it draws streight upwards, whereby the Tongue is the more easily thrust forward out of the Mouth, though in the excessive heat of Fevers, the Fibres of this Muscle are so parcht that the Patient does it with difficulty.

E. Shews this, *Tab.* 8. *Fig.* 3.

Hypsilaglossus

Hypsiloglossus seu Basioglossus.

THis is one of the third pair, it ariseth Fleshy from the Basis of *Os Hyoides*, and is inserted into the middle of the Tongue, according to its longitude, and by drawing it inwards, doth bring it backwards.   *This moves it backwards.*

This being contracted, it brings the Tongue inwards, and backwards.   *Use.*

G. Shews this, *Fig.* 3. *Tab.* 8.

O   Stylogloffus.

Stylogloffus.

This brings the Tongue upwards and inwards.

THis ariseth Fleshy and small, with a sharp beginning from the *Styloidal* process, and growing more broad and Fleshy, is inserted into the lateral part of the Tongue, and it brings it upwards and inwards: it is best found, by discovering of *Styloides* with your Finger, and then your Eye will direct you to it, at the lateral part of the Tongue; in man it is slender, but in Beasts it is double, Fleshy, and thick.

Use.

Its use is thus explained; If either of these Muscles moves, the Tongue is carried either to the right or left side directly, but both moving, its brought back to the *Fauces*.

K. Shews this, *Tab.* 8. *Fig.* 3.

Lingualis.

Lingualis.

This ariseth Fleshy and large from the Basis of the *Os Hyoides*, and so runs according to its longitude, forward to the tip of the Tongue, and is much disputable whether it be a Muscle or not; it's endowed both with oblique, transverse, and right Fibres, all which are so fully sprinkled about the Tongue, that is through its whole Body, that they are scarce divisible, and cannot by the best and most industrious hand be separated. *This moves the Tongue to his Constriction and Dilatation.*

Spigelius doth give these Uses to these pair of Muscles, if they may properly be called so, that the transverse Fibres which are implanted in them do serve to contract the Body of the Tongue and so to thicken it, the oblique dilating it, and separating them from one another, and that the right were framed for bringing it to the Palate and *Fauces* in Constriction. *i.e.*

This Muscle is not to be explained, being disputable whether it be a Muscle or not amongst Anatomists.

Cricothyreoides

Cricothyreoides Anticus.

This moves the Cartilage obliquely downwards.

THis is said to be the first proper pair of the *Larynx*, as is held by *Veslingius* and most Anatomists : it takes its Origination from the fore-part of the Annulary Cartilage, and ends in the sides of the Scutiformal, and hence it gets the name of *Cricothyreoides Anticus*, and is generally reputed to move the Cartilage somewhat obliquely downwards ; it ariseth in the lower and fore-part of the *Larynx*, having a Fleshy beginning ; when this is contracted, it extends the Cartilage *Cricois* or *Annularis*, and so openeth its Cleft for a more deep and greater Voice or Sound.

Use.

F. Shews this at *Tab.* 8. *Fig.* 2. D. D. Shews the same laid bare, *Tab. ead. Fig.* 2. C. Shews the same, *Tab. ead. Fig.* 3.

Æsophægeus

The Explanation of the Eighth Table.

C C. Sternohyoides *at Fig.* 2. *G. G. Shews the same laid bare at Fig.* 3. *ejusd. Tab.*
L. L. Sternothyroides *Fig.* 2. *at N N. Shews the same laid bare Fig.* 3.
M. M. Hyothyroides *Fig.* 2. *at m. m. Shews the same bare at Fig.* 3. *ejusd. Tab.*
E. E. Styloceratohyoides *at Fig.* 2. *F. F. Shewing the same laid bare at Fig.* 3. *ejusd. Tab.*
D. D. Genehyoides *at Fig.* 2. *E. E. Shews the same laid bare at Fig.* 3. *ejusd. Tab.*
C. Miloglossus *shews this at Fig.* 2.
E. Geneoglossus *at Fig.* 2.
I. Ceratoglossus *at Fig.* 1. *D. Shews this also at Fig.* 2.
G. Hypsiloglossus *at Fig.* 2.
K. Styloglossus *at Fig.* 2.
F. Crycoarytenoides Anticus *at Fig.* 1. *D. D. Shews the same laid bare*

Æsopha-

TAB. VIII   Fol. 56.

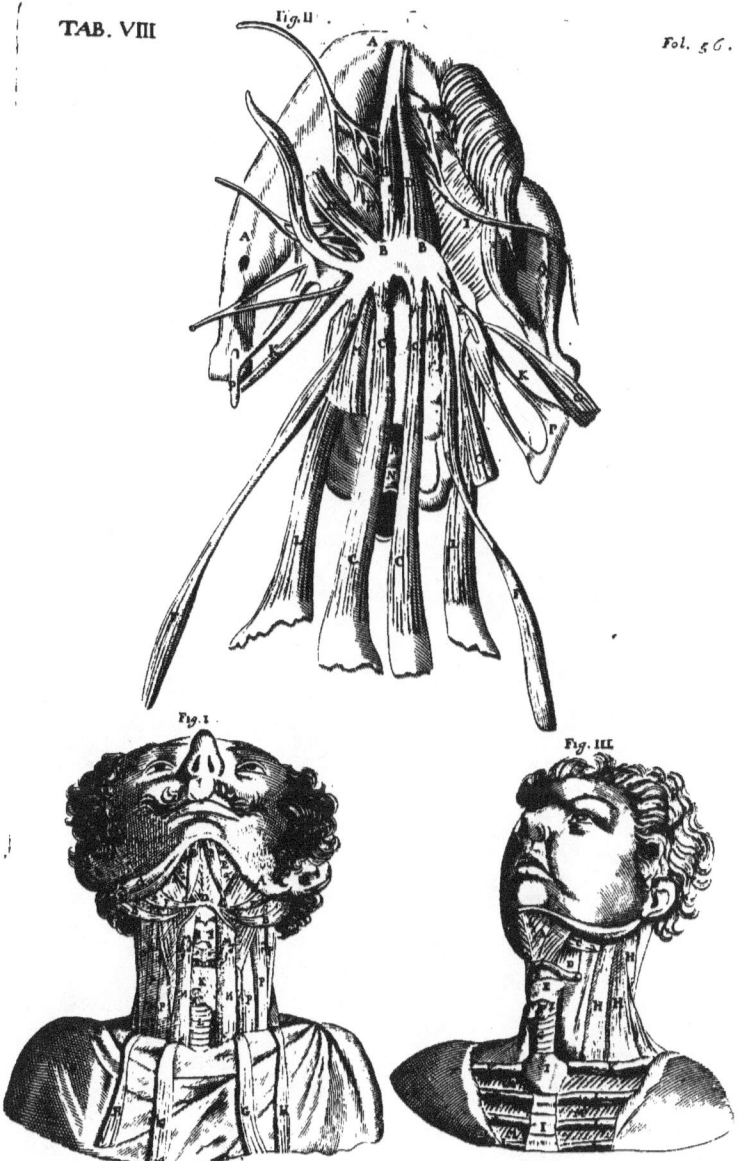

Æfophageus feu Sphincter gulæ.

The *Æfophagus* is a round Channel, by and through which our Nourishment, as both Meat and Drink doth pass from the Mouth into the Stomach, and this Action of Deglutition is performed by the Fibres of the *Æfophagus*, and its Muscles; for whereas we Eat and Drink at our wills and when our pleafure is, this Action is performed when thus made ufe of, by fuch Inftruments as do ferve for a voluntary Motion, as Mufcles, which are to be numbred with their Animal Actions: and although it may ferve for a Natural ufe, (as for Nutrition) yet it is no lefs Animal than Breathing, the which alfo is cuftomary to Nutrition, yet is it Animal ; Now this *Sphincter Gulæ* is very officious in this fervice, for when a due Maftication is made, this by its Conftriction doth drive it downwards. *This contracts the Fauces. Its Ufe.*

It arifeth Fleshy, and is a broad Mufcle wholly encompaffing both the fore and back part of the *Æfophagus*, arifing laterally according to the length of the Scutiformal Cartilage, encompaffing the *Æfophagus* like a *Sphincter*; This Mufcle being carefully raifed, you will much better find *Stylopharyngæus & Cephalopharyngæus*.

D. D. Shews this, *Tab. 9. Fig.* 5.

Stylopharyngæus.

This dilates and opens the Æsophagus.

THis third pair is so called, and do arise with a small beginning from the inner part of the *Styloidal* process, and descending with its thin body, is implanted by a Membranous Tendon into the *Thyrois*, the sides of the *Os Hyoides*, and the Root of the Tongue; this helpeth forwards the former in its Contraction.

Use.

Whereas there are appointed two as Dilators of the *Æsophagus*, so this is accounted as one of the Constrictors, though *Veslingius* thinks it rather Dilates.

C. C. Shews this, *Tab. 9. Fig. 5*.

Cephalopharyngæus.

THis ariseth from the *Cranium*, and the first Vertèbre of the Neck where they are joyned, and so descending, is implanted into the lateral sides of the *Os Hyoides*, *Cartilago Scutiformis*, and the beginning of the *Æsophagus*, for whom he seems to make a Coat; and by lifting this up doth constringe the *Fauces* in the Deglutition. *This doth constringe Fauces.* *Use.*

A. A. Shews this, *Tab.* 9. *Fig.* 5.

Cricoary-

Cricoarytænoides Posticus.

This extends the Larynx.

BY *Spigelius* this is reckoned as one of the first pair of the *Larynx*, extending or opening its Cleft, and by *Casserius* it is called *Par Cucullare*, it ariseth Fleshy from the hinder and lower part of the *Cricoides*, and running up with right Fibres, and repleating the Cavity of the *Cricoides*, is Nervously implanted into the inner seat of the *Arytænoides*, and by division of the two *Arytænoidal* Cartilages, the *Larynx* is opened. This doth extend the *Arytænois*, and by converting it backwards to the outward parts, it opens the Epiglot.

Use.

D. D. Shews this at *Tab. 9. Fig. 1.* G. G. Shews the same laid bare, *Tab. ejusd. Fig. 2.* C. C. Shews the same, *Fig. 4. ejusd. Tab.*

Cricoary-

Cricoarytænoides Lateralis.

This is generally allowed one of the third pair, extending the *Larynx* to the fide, or opening its fecond *Rima* or Cleft, it arifeth from the lower part of the Annulary Cartilage upwards, and is inferted into the lateral external part of the *Arytenoides*, opening the *Larynx* by the oblique diduction of its Cartilages. *This extends it obliquely lateral.*   *Ufe.*

Here's alfo obfervable; That by how much the fecond pair of the Contractors doth draw to a mutual Contract, the fecond pair of the Extenders do bring them outwards, and fo open them. This is generally allowed to extend the *Larynx* laterally, and fo doth open the *Rimula*.

E E. Shews thefe laid bare, *Tab. 9. Fig. 2. C.* Shews the fame *in fitu*; *Tab. ead. Fig. 3. C.C.* Shews them *in fitu*, *Tab. ead. Fig. 5.*

Arytænoides.

Arytænoides.

This contracts it obliquely lateral.

THis is also called *Claudens Secundum*, its very small and Fleshy, and ariseth with oblique Fibres from the *Arytænois*, where it is joyned to the *Cricois*, and is again inserted into it, where it connecteth its self with its Companion;

Use. This moveth the *Arytænois* obliquely, and to either side, and so by constringing its Basis, doth shut the *Glottis*

This is called *Arytænoides* or *Guttalis*, and whereas there are allowed two motions of the *Larynx*, by which it is either contracted or dilated, shut or opened, so doth both these Dilatations and Constrictions, or Clausion and Apertion proceed from their proper Muscles. Thus when the *Thyrois* is dilated, the *Arytænoides* is shut; and thus according to *Galens* opinion, the *Larynx* is contracted when the sides of the *Thyrois* or *Scutiformis* are contracted and moved inward, dilated when they are extended and brought outwards, and shut when the *Arytænois* is constringed and brought into a cavity, opened when it is brought outwards and extended.

The action of these Muscles are most remarkable, when we forcibly stop our Breath for some time, for then it prevails against the contrary endeavour of the Muscles of the *Thorax*, which serve to Respiration, and shuts the *Arytænoides* so close, that no Air can enter in.

F. F. Shews the same laid bare, *Fig.* 2. *ejusd. Tab. B. B* Shews them at *Fig.* 4. *ejusd. Tab.*

Thyreoary-

Thyreoarytenoides.

This is one of the fourth pair, arising Fleshy, broad, and is transversely implanted in the Cavity of the *Larynx*, and from the middle inner part of the *Thyrois*, and being carried upwards according to its length, is inserted into the lateral part of the *Arytænoidis*, which makes the *Glottis*, the which constringing doth shut the *Larynx*; This Muscle is best found by dividing *Cartilago Thyroidis*, from the *Cricoidis*, *Arytænoidis*, and subjacent Muscles, the Coats of them being carefully preserved, after which will plainly appear this. *This contracts is directly.*

This pair if they be inflamed, in a Squinancy, when as they do exactly shut the *Rima* or Cleft, it brings Death along with it. *Obs.*

D. Shews this, *Fig* 3. *Tab.* 9. *B. B.* Shews the same at *Fig.* 6. *ejusd. Tab.*

Sphenopharyngæus Primus.

This doth dilate the Fauces.

THe Muscles of the *Fauces*, by some the *Pharynx* or beginning of the *Æsophagus*, are those which do serve for Deglutition, and therefore Nature hath planted at the top of the *Æsophagus* Muscles here as Instruments of voluntary motion for acting to our wills or pleasures; and as some of these do serve for Constriction, so also are others as useful for Dilatation, amongst which are these reckoned as the first pair.

This ariseth thin and Nervous nigh the sharp Appendix of the *Os Cuneiforme*, descending by the inward Cavity of the *Pterygoides*, and is inserted by a small Tendon into that Skinny part of the Pallate, from which the *Gargareon* seems to proceed, and doth dilate these parts for Reception of their Nutriment.

Neither this nor its fellow are rais'd well or distinctly without much trouble and difficulty.

B. B. Shews this at *Tab.* 9. *Fig.* 4.

Spix.

The Explanation of the Ninth Table.

D D. Æsophagæus *at Fig.* 4.
C. C. Stylopharyngæus *at Fig.* 4.
A. A. Cephalopharyngæi *at Fig.* 4.
B. B. Sphenopharyngæi *at Fig.* 4.
D. D. Cricoarytenoideus Pofticus *at Fig.* 1. G. G. *Shews the fame laid bare at Fig.* 2. D. D. *Shews the fame at Fig.* 5.
E. E. Cricoarytenoideus Lateralis *at Fig.* 2. C. C. *Shews the fame laid bare at Fig.* 3. C. C. *Shews the fame alfo at Fig.* 6.
F. F. Aritenoides *shews this laid bare at Fig.* 2.
D. Thyroaritenoides *at Fig.* 3. B. B. *Shews the fame at Fig.* 6.

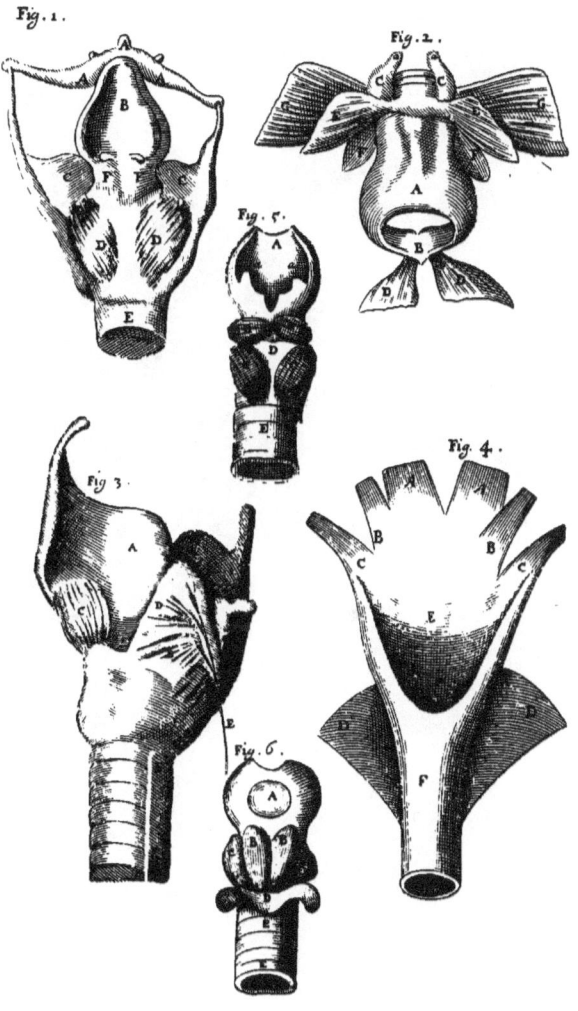

Sphenopharyngæus Secundus.

THis second is by some allowed to arise from the same Origination as the former, and is inserted into the lateral and back part of the *Fauces* and *Æsophagus*, the which drawing downwards, they do dilate the Cavity of the *Fauces* and *Gula*; To find these after you have raised the *Larynx* and *Æsophagus*, leave the *Fauces* entire, then divide the *Fauces* themselves from *Os Palati*, till you come into the Cavity; then carry your Knife close internally to the *Os Cuneiforme*, and being thus divided, you will with ease find both their Originations, and with as much pleasure you may dissect them.

This helps the former in its dilatation.

Use.

This also is shewn at *Tab.* 10. *Fig.* 1. at *B. B.*

Next to these should I have put in the Stomach and Intestines, but Dr. Willis hath given so full an Account of them, and shewn their Fibres so exactly, that whoever desires to take a view of them may be very well satisfied thereof in his Book named Pharmaceutice Rationalis.

Pterygoides

Pterygoides Externus five Abducens.

This brings the Mandible forwards.

THe lower Mandible hath allowed it variety of Motions, and these are both very necessary to Human life, as also for well ordering of their Designs: for how can the Meat be either received into the Mouth, or there chewed or lessened, unless the Mouth were opened, and then shut again; the Teeth do act their parts, and by a mutual Collision and Comminution towards a fair Deglutition, do make a happy progress towards the health of Man; without the benefit of Muscles, the Grinders must lie still, and the Incisors grow dull; the Mandible hath its variety of Muscles granted it for the discharge of its variety of Offices, amongst which this is accounted as one of its fourth pair, and doth arise from the External part of the *Processus Aliformis* in part, as also from the rough and sharp Line of the *Os Sphænoidis*, strong, Nervous, and Fleshy, and so marching down large, is inserted by a strong broad Tendon into the inner part of the lower Mandible, laterally, just under the Tendon of the Temporal Muscle, and

Use. doth move the Mandible forwards, and as it were doth abduce it from the Head.

E. E. Shews this at *Tab.* 10. *Fig.* 1.

Pterygoides

Pterygoides Internus seu Adducens.

This ariseth thick and short from the inner Cavity of the *Processus aliformis*, being first Nervous, then Fleshy, and is inserted by a broad, long, and Nervous Tendon into the lower Mandible internally lateral, the which drawing it upwards, doth help the Office and Action of *Temporalis*, bringing the Mandible inwards and backwards. These two Muscles do not appear until the whole dissection of the Tongue, *Larynx*, and *Gula* be fully compleated.

This brings it backward.

Use.

D. D. Shews this at *Tab.* 10. *Fig.* 1.

S   Longus.

Longus.

It contracts the Neck.

THe Neck is as an Appendix to the middle Venter, or a middle between the Head and Trunck. Such Animals are destitute hereof, which do not move their Heads with the Trunck of their Bodies, as Fishes, Frogs, &c. This Neck hath a four fold motion granted it, as forwards, backwards, and to either side, and every of these Motions are performed by the benefit of Muscles, of which some are called Flectors, others Extensors; of the Flectors, the first pair are called *Longi*, lying under the *Æsophagus*.

This Muscle ariseth sharp and Fleshy from the forepart of the Body, from the fifth and sixth Vertebre of the *Thorax*, where the Rib joyns its self to him, and so running up under the *Æsophagus*, is joyned to the sides of the Bodies of all the Vertebres, ascending until he comes to the first of them, where meeting with *Scalenus*, they insert themselves by a sharp Nervous Tendon into the transverse process of the first Vertebre of the Neck.

Use. The Neck by the benefit of these with the Head, is bent or contracted forwards, one only operating, it carrieth it to the sides

A. A. Shews this, *Tab.* 10. *Fig.* 2.

Scalenus

The Explanation of the Tenth Table.

FIG. I.

A *A.* Temporalis *laid bare.*
B. B. Maſſeter.
C. C. Digaſtricus *or* Biventer.
D. D. Pterygoideus Internus.
E. E. Pterygoideus Externus.
F. Quadratus Riolani.

FIG. II.

r. ſ. t. v. Os Ptery oides.
O. O. Muſculus Pterygopalatinus *which depreſſeth the* Uvula, *and with it the Glandule of the Palate.*
X. X. *Shews its Tendon which is reflected about the neck of the* Os Pterygoides *as on a* Trochlea, *and is inſerted into the Glandule of the Palate.*
r. *Shews the neck of the* Os Pterygoides *with its ſmall head.*
B. *Shews the* Glandula Palati.
d. *The* Uvula.
a. a. *Part of the Muſcle* Pterygoideus Internus *to which is adjacent* Pterygopalatinus *mentioned at* O. O.
C. *Shews its Origination ariſing from the lower part of the Cavity of the* Os Pterygoides.
q. *Shews another Muſcle called* Sphenopalatinus *which attols the* Glandula Palati, *and with that the* Uvula.
f. f. *Shews its Tendon implanted in the ſide of the* Glandula Palati.
Z *Shews its Origination out of the* Os Sphenoides.
g g *The Interior Cavities of the Noſtrils*

  Moreover as touching theſe two Muſcles (beſides what hath already been ſaid as to their Uſes) this Obſervation is very material as to their Uſefulneſs, (viz) in Hauking *or forcibly raiſing up any tough* Flegm *or* Lapidouſe *Matter lodged about the* Æſophagus, *theſe are of very great uſe as to the diſpatch of the ſame forward, and ſending it outwards, being here planted as two ſtrong Ligaments which can give force to the* Pallates *raiſing it ſelf for the diſcharge of the ſame*

FIG. III.

A Muſculi Longi.
B. Scalenus.
C. Maſtoideus.

Scalenus

Tab. 10. fol: 71.

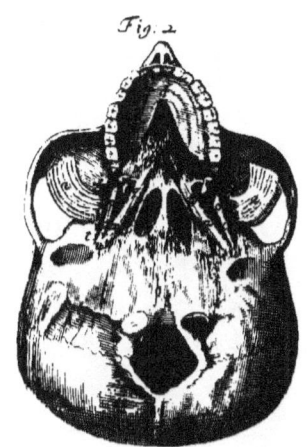

Fig. 2.

Fig: 1. Fig: 3.

Scalenus five Triangularis.

THese pair of Mufcles refembling a Triangular Figure, by fome Anatomifts are not improperly called *Triangulares*. *This contracts the Neck as the former.*

This arifeth from the firft and uppermoft Rib of the *Thorax*, broad and Flefhy, and then narrowing himfelf, in his upper courfe he beftows tranfverfe Fibres upon all the tranfverfe proceffes of the Neck, and is inferted as the former, and doth alfo help the former forward in their motion ; This Mufcle hath a peculiar Cavity allowed it, through which the Arteries defcending to the Arm, and the Veins afcending from thence do pafs.

If thefe Mufcles do work together, they do contract and bring the Neck and alfo the Head right forwards, but if only one do operate, it inclines the Neck to one fide forwards. *Ufe.*

B. Shews this at *Tab.* 10. *Fig.* 2.

Pectoralis.

Pectoralis.

It moves the Arm forwards.

THe upper part of the Arm reaching from the top to the Cubite, is that which we generally do call the Arm, and this is moved by several Muscles, the first of which is called either *Adducens Humerum*, or *Pectoralis*, by some the Boxing Muscle, it possesseth the forepart of the *Thorax*, and ariseth first from the middle of the *Clavicle*, where he is most round, and next the Breast Fleshy, then from the *Sternon* according to his length, and is annexed to his Cartilages; Thirdly, from the Cartilages of the sixth, seventh, and eighth Ribs, and is there Nervous in his Origination, and so proceeds Fleshy and large until he arrives at the lateral part of the *Thorax*, where he is converted into a short, broad, and strong Tendon, and is inserted into *Os Humeri*, and according to his length between *Deltois* and *Biceps*.

Use.
This bringeth the Arm to the Breast forwards, and hence by some hath it given it the name of the Boxing Muscle, and this either directly, or somewhat upwards, or downwards, according to the contraction of its upper, middle, or lower Fibres.

B. Shews this in *Tab.* 16. *c. b. d.* Shew its diverse Fibres.

Subclavius.

Subclavius.

This is called the *Subclavius* which is seated under the *Clavitle*, and lodgeth between the first Rib of the *Thorax*; This is accounted the first Muscle of the *Thorax*, it ariseth from the Inferior part of the *Clavicle*, and being enlarged forwards with Oblique and Transverse Fibres, it is implanted into the first Rib near the *Sternon*, and by drawing it upwards and outwards, doth dilate the *Thorax*. *This brings the Clavicle upwards and outwards, and doth open the Thorax.*

The use of this Muscle is for drawing down the *Clavicle*, when it is moved upwards with the *Scapula*, for when the *Scapula's* are attolled, the *Clavicles* are lifted up with them, the which to reduce into their proper places, the *Subclavius* is to be brought downwards; and hence in Fractures of the *Clavicles*, if they be fractured near the *Sternon*, the Arm with the *Scapula* does soon fall downwards, and that part which is next the *Sternon* doth ascend and is raised upwards, as *Hippocrates* doth observe, **Lib. de Fractur.** and on the contrary, if they be broken near the *Acromium*, you will find neither part to ascend: the cause of which, the same worthy Anthor doth fortifie with this ingenious Reason; Whereas this Muscle is inserted to that part which is next the *Acromium*, when the *Clavicle* is fractured two ways, this *Subclavian* Muscle is presently contracted; and thus the Arm draws the dependent *Scapula* together with its self, whilst the other part is detained in its place by the strength of the Muscle. *Use and*

F. Shews this, *Tab.* 11.

T   Serratus

Serratus Major Anticus.

This brings the Scapula forwards.

THese Muscles of the *Scapula* have not their Names given them from their Use, but rather from their Figures, as this pair much resembling the Teeth of a Saw, and hence called *Serrati*. This is the second Muscle placed in the side of the *Thorax*, arising from the third, fourth fifth, sixth, seventh, eighth, ninth, aud tenth Ribs, with a Saw-like, Nervous, and sharp Origination, hence growing Fleshy upon the Ribs, he ascends and inserts himself Fleshy into the whole Basis of the *Scapula* inwards; It is much Disputable both by *Spigelius* and *Veslingius*, whether the Origination and Insertion be not *è contra*; the motion of this Muscle is promoted by the Oblique Descendent Muscle of the *Abdomen:* when this is contracted to its beginning, it draws the *Scapula* forwards, as also the *Serratus Minor*, but somewhat downwards.

Use.

And also as I apprehend, another Use may be to tie or fasten the *Scapula* to the Breast.

A. Shews this at *Tab.* 16. *a. a. a.* Shews its Origination and its Tendon, *D. D.* Shews the same at *Tab.* 11. *F. F. F. F.* Shews the same in its place in *Tab.* 1.

Serratus

Serratus Minor Anticus.

This is the fourth, which wholly lies under the *Pectoral* Muscle in the fore part of the *Thorax*, and is called *Serratus* from its being like a Saw, acuminated with various Fleshy Fibres; *Minor*, as touching its difference with the *Major*, it is substrated to the *Pectoral* Muscle, arising from the four first upper Ribs (but one) by so many serrated Originations; and so descending towards the *Scapula* narrows himself, and is inserted broad, Fleshy, and Nervous into the *Processus rostriformis* of it, and draweth it forward to the *Thorax* : if they work both together, they do bring the *Scapula* to the Breast : if the upper they bring it upwards, and *è contra* : if the lower, they bring it downwards. *This brings it upwards.* *Use.*

This Muscle hath also (as I conceive) the same Use with the former, binding the *Scapula* forward to the Trunck of the Body.

F. Shews this laid bare, *Tab.* 11.

Intercostales

Intercostales Externi.

These do dilate the Thorax.

These have their Originations from the Transverse processes of the Back, where the Ribs are joyned, and so proceeding Fleshy, do fix themselves all along from the lowermost part of the upper Rib, external to the upper part of the lower Rib, and so doth proceed to the Cartilage of the *Sternon*:

Use. The outward Muscles being contracted, do draw the Ribs towards their Originations upwards and outwards, in Respiration; hence the *Thorax* is dilated, and so its Cavity is made wider.

Obs. All these Muscles are endowed with Oblique Fibres, and these intersecting each other, do shew their differences from the *Interni*. As a good Observation to the young Chirurgeon, let him take care in opening of any Abscesses or Empyemas which may happen here, for if he makes a direct Incision here downwards, he cuts and divides all the *Spermatick* Fibres, and therefore in these cases he is advised to make his Incision in an oblique manner.

K.K.K.K. Shews this at *Tab.* 11. O.O.O. Shews the same, *Tab.* 5.

The Explanation of the Eleventh Table.

F Subclavius.
D. D. Serratus Major Anticus.
E. Serratus Minor Anticus *shewing it laid bare.*
K. K. K. K. Intercostales Externi, Intercostales Interni.
C. C. Subscapularis.

Fol. 70. TAB. XI.

Intercostales Interni.

These have their Originations where the Ribs do begin to bend inwards, and run from the lower to the upper part of the Ribs, not only to the Cartilage, but under that to the *Sternon*; these Muscles do work contrary to the former, for these do bring the *Thorax* downwards and inwards in Expiration, whence it becomes Constrict, and the Cavity is made less. *These do move the Thorax and constringe it.*

Moreover, whereas the External Muscles do end about the beginning of the Cartilages, so as that there are left Intercartilaginous spaces: hence is it, that Nature, that provident Mistress, who abhors all vacancies, hath filled all these empty spaces with these Internal Muscles, and hath raised the same to the Exteriour Superficies of its space or those spaces.

These you may also see with the former.

V   Pectoralis

Pectoralis Internus seu Triangularis.

This doth constringe the Thorax.

THis by some is accounted the sixth Muscle of the *Thorax*; it is a small and thin Muscle arising from the inner part of the *Sternon*, and adheres to the Cartilage of the upper Ribs, sending forth on either side four small Projections to the Osseal Extremities, by which the third, fourth, fifth, and sixth true Ribs are joyned to the Cartilages, by the adduction of which, they do constringe the *Thorax*, and do somewhat depress it forward.

Use.

This Muscle is not to be shewn by any Figure.

Cremasteres

Cremasteres sive Suspensorii.

There are three proper Coats allowed the Testicles, 1st. *Erythroides* or *Tunica rubra*, and 2dly. *Elytrois*, or *Vaginalis*, and 3dly. *Tunica Albuginea*, or *Nervea* : to the External Membrane of the first are adnated the Cremaster Muscles, one on either side, the which in Men have their Originations from that Ligament which is in the *Os Pubis*; in Dogs and other Animals they are seen to take their Originations from the Tendons of the Transverse Muscles, and their Fleshy Fibres do run through the whole length of the Vaginal Coat, especially in its back part; for which cause the outward Superfices of this Coat is seen to be Asperate and Fibrous, the inward smooth, and covered with a certain waterish Humidity, and is strongly annexed to the lower part of the Testicles. *These keep the Testicles from falling down.*

Regnerus de Graaf doth allow a three fold Use of this Muscle; as first that it keeps the Testicles from Cold; Then that it keeps them up from falling down, the which by their weight, should it once so happen, they would hinder the *Spermatick* Vessels in their Operations; And lastly, as various Authors have writ and observed, That they attract them for a better Excretion of the Seed, as is seen in the Act of Venery. *Bartholinus* doth witness, That there are such who have this so strong, that they can according to their will retract the Testicles, and then again dismiss them. *Use.*

Tab. 13. *Fig.* 4. at *C. C.* you have this, *D. D.* Shews their Fleshy Fibres.

Erector

Erector Penis, sive Collateralis.

T is is said to erect the Penis.

THe Seed made, prepared, and elaborated in the *Spermatick* Vessels do require a proper Instrument for its discharge into that part which Nature at first designed it for, by which means we see the like produced by the help of this Instrument. *Plato in Timæo* did suppose the *Penis* to be some certain Animal, which could produce such strange effects as touching both Generation and Propagation, but although it is no Animal, yet it must properly be allowed an Animal-part and Instrument: Its placed in the lower part of the Belly, for the more commodious executing its Office, it takes its Original from a strong Foundation, as from the Bones of the *Pubis*, to whose Root it is most firmly planted; we pass by its Figure and Substance, and come to its Muscles.

This Muscle has his Original from the Appendix of the *Coxendix*, beneath the beginning of the two Nervous Bodies, in whose Interior part their thickest Fibres do terminate and vanish.

Use.

Spigelius doth assert that they take their names from their qualities, and that they do erect the *Penis*, and in coition do preserve the same; but this is denied by *Regnerus de Graaf*, as you will see in the next Chapter: for these Muscles rather depress the *Penis*, that so the Seed may be the more straightly ejaculated into the *Uterus*.

S. S. Shews this at *Tab.* 13. *Fig.* 1. *T. T.* Shews the same, *Tab. ead. Fig.* 2. G. G. Shews this, *Tab.* 12.

Accelerator

The Explanation of the Twelfth Table.

G *G.* Erector Penis.
F. F. Accelerator Penis.
K. K. Levatores Ani.
I. Sphincter Ani.

Accele-

FOL: 79.   TAB XII.

Accelerator Penis.

Besides the former, the Virile Member hath two other Muscles allowed it called *Urethram Trahentes*, arising Fleshy from the *Sphincter Ani*, and joyning with its partner internally lateral, and marching by the fore part of the *Penis*, is inserted into the *Urethra*, and is generally asserted that it was framed for the dilating of the *Urethra*; but the Use thus designed these Muscles is much rejected by *Re n. de Graaf*, the which he confuteth by these Reasons, That when all Muscles do work in their own proper method, their Venters do tumefie, and their ends do approach nearer each other; the which being granted, it cannot thus happen that the *Penis* should be extended, the Action of the Muscle being Contraction, and this being most contrary to Extention; neither can the *Penis* obtain Erection by the benefit or help of these Muscles, for it rather would appear depressed than erected these being contracted, and they being planted in the lower part, or under the *Penis*, taking their Origination from the Appendix of the *Coxendix*, and so implanted to the lower part of the Yard.

As to the two also which are assigned by other Authors to dilate the *Urethra*, they are in no ways capable to perform this, these Muscles running through the middle of it, are firmly united to each other by one extremity of Fibres, whilst the opposite to the former obliquely running over the *Urethra*, do send the same into the sides of its Nervous Bodies; but as to the Erection of the *Penis* two kinds of Vessels do seem chiefly to serve with the Muscles for the performance of this, as Nerves and Arteries; but of these you may plentifully satisfie your self in *Regner. de Graaf, de Organ. Viror. Fol.* 154. &c.

This you have at *R. R. Tab.* 13. *Fig.* 1. *Fig.* 2. at *S. S.* you have the same, *ead. Tab. F. F.* Shews the same, *Tab.* 12.

This said to dilate the Urethra.

X   Musculi

Musculi Clitoridis.

This extends the Clitoris.

THe *Clitoris* hath variety of names bestowed on it, as *Amoris Dulcedo*, *Oestrum Veneris*, *Libidinis Sedes*, &c. it differs from the Virile Member if you consider its whole Fabrick; First, because its bifurcated parts are twice longer when joyned, in the *Penis* when the parts are conjoyned they are four times longer than the bifurcated parts, Then because it hath no such like Channel as hath the *Penis*, neither is its Glans perforated as is that of the *Penis* in Men; All Anatomists have by consent allowed that there are Muscles annexed to the *Clitoris*, but as touching the number of them there is held a Dispute; we judge and allow of two arising from the Bones of the *Coxendix*, and running above its *Crura*, are implanted in them; This Muscle we call *Graafiani*, and by this the *Clitoris* is raised.

Obs. This Use we think proper to ascribe to the *Clitoris* and its Muscles, by the Contraction of the *Clitoris* and compressing its Thighs, it doth distend the third Body much more with the Glans.

There is also another pair of Muscles ascribed to the *Clitoris* by Authors, arising from the *Sphincterani* with a broad beginning, passing by the Lips of the *Pudendum*, between the *Clitoris* and *Plexus Retiformis*, and it is so annexed to the *Clitoris*, that it is judged that it was appointed rather for the contracting of the Vaginal Orifice, than for erecting the *Clitoris*; and from hence we presume to call it *Musculus Labiorum Uteri Contractor*.

Use.

C. C. Shews these at *Tab.* 13. *Fig.* 4. by this Muscle the *Clitoris* is depressed, *D. D.* Shews its Fleshy Fibres, *E. E.* The Fleshy Fibres of the *Sphincter* annexed to the Nervous substance of the *Clitoris*.

Levatores

Levatores Ani.

THat part we generally call the *Anus* is that which is the end of the *Inteſtinum Rectum*; theſe ariſe from the li- *This lifts it up.* gaments of the *Coxendix*, and *Os Sacrum*, under the Bladder, ſmall, thin, broad, and Membranous, and are inſerted into each ſide of the *Podex*; Theſe Muſcles being very much relaxed, do ſuffer a *Procidentia Ani*, or rather *Prolapſus Inteſtini*, *Uſe.* theſe are beſt diſcovered before you remove the *Inteſtinum Rectum*, *Veſica*, and adjacent parts.

R. R. Shews theſe at *Tab.* 13. *Fig.* 2. K. K. Shews the ſame, *Tab.* 12.

Sphincter

Sphincter Ani.

This purses up the Anus.

THis from its Use is called *Constrictor Ani*, or *Orbicularis*, it ariseth from the lower Vertebres of the *Os Sacrum*, round, and broad, joyning himself largely circular to the *Inestinum Rectum* with Transverse Fibres much thicker above than below, where he adheres so firmly to the *Cutis*, that it is scarce separable: and hence by some Anatomists it is called *Cuticulosus*.

Use. We acknowledge the Use and Nobility of either of these, for when they any wise suffer a *Paralysis*, this being hurt, the Excrements involuntarily do come down, and for the service they do in this case, they are called *Constrictores*

Q. Shews this at *Tab.* 13. *Fig.* 2. *I.* Shews the same, *Tab* 12.

Sphincter Vesicæ.

THe Urinary Bladder is an Organick Membranous part *This parts* of the lower Belly, the which is formed as a receiver *the Bladder.* to take the Urine into it, or Serum which passeth from the Kidneys into it, and at due time doth also serve for a discharge of the same, and so this Bladder hath two Muscles given it, the one is said to retain the Urine in it, the other drives it out; How unkind would Human life be, if it should be continually attended with a continual dropping of Urine, as also how troublesome would it be to Mankind to be perplexed with as great a stoppage; therefore kind Nature as she hath been so free as to give the Bladder one pair to keep it in whilst it is convenient to discharge it, so also hath she been as provident to let it out when the Bounds of Nature commands a discharge thereof: the first from its Office hath gained the name of *Sphincter*.

This is placed orbicularly in the neck of the Bladder, scarcely distinct from the substance of it, only where you find a round Induration, this is the same, this doth keep the Urine from involuntary falling out from the Bladder, pursing it up.

O. O. Shews this at *Tab.* 13. *Fig.* 1.

Detrufor Urinæ.

This lets it out.

THis is faid to arife betwen the common, and fecond proper Coat of the Bladder, the firſt Coat properly being his, if not him, and therefore it is very Fabulous, which ſome Phyſitians fo ſtrongly contend for, that beſides the former they will allow the Bladder many other Muſcles, but in truth the beſt of Authority doth not afford any other of the Bladder beſides theſe two already named.

This Muſcle therefore is only the middle Coat of the Bladder, which conſiſting of Carnous Fibres running length-ways ſerves to the expulſion of the Urine. The tone of theſe Fibres is much injured when the Bladder is overmuch diſtended with Urine, or when it is held too long.

M. M. M. Shews this, *Tab.* 13. *Fig.* 1.

Diaphragma.

The Explanation of the Thirteenth Table.

C D. Cremasters *at Fig.* 3.
S. S. Erector Penis *at Fig.* 1.
T. T. *Shews the same laid bare at* Fig. 2.
R. R. Accelerator Penis *at Fig.* 1.
S. S. *Shews the same laid bare at Fig.* 2.
C. C. Musculi Clitoridis *at Fig.* 4. *and at* D. D. E. E.
R. R. Levatores Ani *at Fig.* 2.
Q. Sphincter Ani *at Fig.* 2.
O. Sphincter Vesicæ *at Fig.* 1.
M. M. M. Detrusor Urinæ *at Fig.* 1.

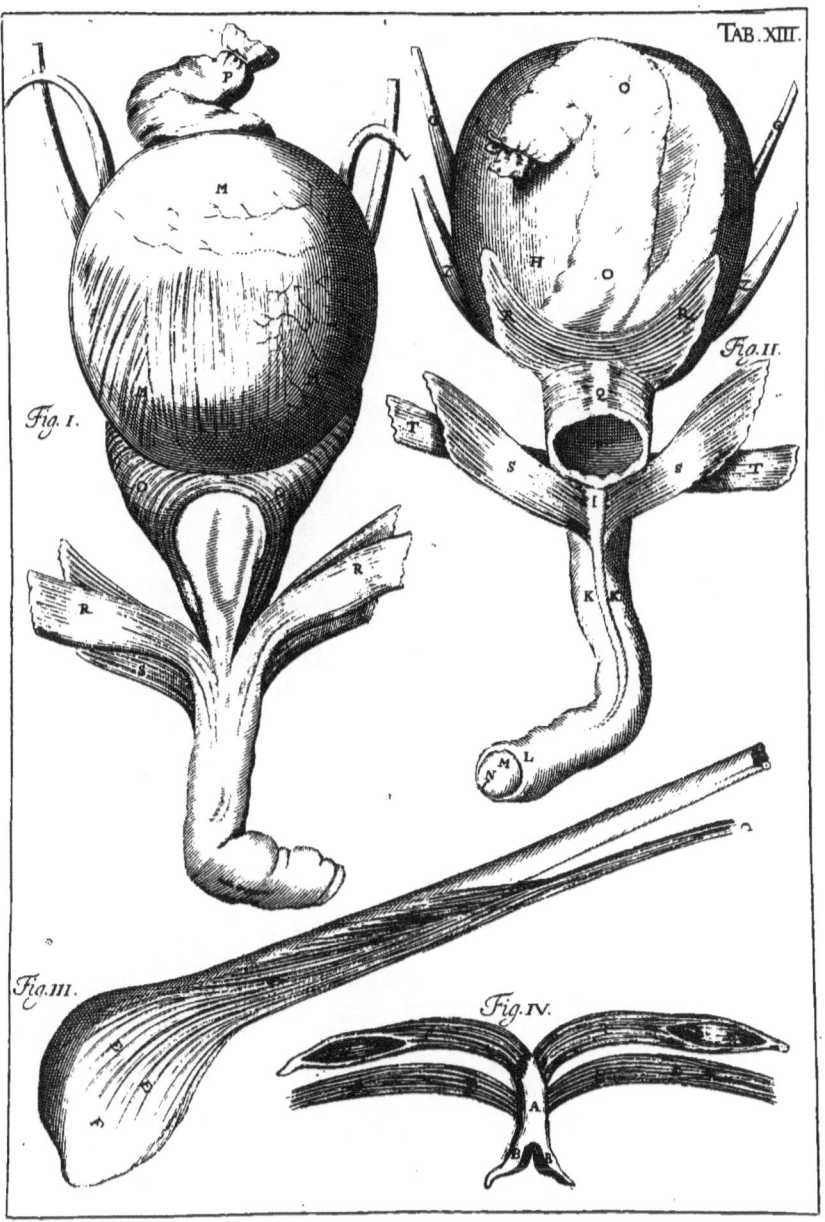

Diaphragma.

THis Muscle hath its Origination from the first Vertebre of the Loins, obtaining a circular Figure, and much different *in situ* from other Muscles, answering in magnitude the transverse bottom of the *Thorax*, and is inserted into the forepart of the *Sternon*, and Termination of the Trise Ribs, and to the twelfth Rib, as also to the extremities of the bastard Ribs, on the Sides. *This distinguishes the lower Belly from the middle.*

It is in its middle (for its greater strength) Membranous and Nervous, to which middle the Fleshy Fibres do run as to their Center: Wounds happening in this Center, are reputed Mortal, because hence suddainly doth follow a present decay of Respiration, and very troublesome Convulsions, whereas Wounds happening in the Fleshy parts hereof, are void of this danger allowed by *Galen*, and this is confirmed by various examples. *Off.*

The *Diaphragma* borrows its Origine from the Vertebres of the Back near the Loins, and round the Termination of the Ribs and *Ensiformis Cartilago*, and hath its Tendon in its Center or middle of it: and by contracting it self, moveth downward, bringing it self from an Arch toward a Plain, and so enlargeth the Cavity of the *Thorax* to give way to the blown up *Lungs*, inspired with numerous Particles of Air. *Use.*

> Next to this should I have Discoursed of the Heart, that Royal Muscle of the Body, by the Reciprocal motion of whose Fibres all that Blood is let both into it and out of it, by which our Human Pile is kept up and preserved: but Doctor Lower having already so fully written hereof, and also given such exact Figures of all its variety of Fibres, I recommend the Reader wholly to his Book De Corde, my task more properly relating to those of the Artus.

This you have at *Tab.* 16. *Fig.* 2.

Here

Here let the Body be turned upon its Face.

Cucullaris five Trapezius.

This mov's the Scapula variously.

THis is the first, the which with its Companion doth very aptly express a Monks Hood : it takes its Origination Fleshy from the lower part of the *Os Occipitis,* and from the Spines of the Vertebres of the Neck, and the eight upper Spines of the *Thorax,* and springeth Membranous, broad, and running externally towards the *Scapula,* grows narrower, and is inserted into the whole Spine of the *Scapula,* and near half his Basis, as also to part of the *Clavicle,* by a broad, Nervous, and Fleshy Origination, and by the variety of Fibres al-

Use.

lowed it, it is variously moved, as upwards, downwards, directly, obliquely, according as its Fibres are variously contracted ; Divide this Muscle from its partner at their Originations from the Spines of the Vertebres, and being so followed and cleared from the *Os Occipitis,* the Muscles underneath this will much better appear.

And I conceive another Use of this Muscle may be to fasten the *Scapula* to the Vertebres of the Neck and *Thorax*; but the chief Use of it is to move the *Scapula* obliquely upwards.

This you have at *A. A. A. B. Tab.* 14. *B.* Shews its Tendinous Insertion into the *Scapula, a. b. c.* Its three sorts of Fibres.

Latissimus

Latisſimus Dorſi, ſive Aniſcalptor.

This *Abducent* Muſcle, or *Latiſſimus* is ſo called from its magnitude, the which with its Companion doth near cover the Back, it ariſeth with a large Membranous beginning from the Spines of the Vertebres of the *Thorax*, between the *Os Sacrum*, and the ſixth Vertebre of the *Thorax*, as alſo from the upper part of the *Os Ileon*: his Origination here is chiefly Membranous, but running higher, ſo ſoon as it attains the curvation of the Ribs, he grows Fleſhy, and in his aſcenſion becoming narrower; is carried over the lower Angle of the *Scapula*, and by a ſtrong and ſhort, but broad Tendon, he is implanted below the upper head of *Os Humeri*, between the *Pectoralis* and *Rotundus*, great care muſt be had, leſt in the raiſing this Muſcle from his Origination, you do take up the Origination of the ſubjacent Muſcle *Serratus Major Poſticus*, and if you be not very careful in your diſſection, you will borrow from *Quadratus Lumbi*, as you raiſe him from the *Ileon*, to which he firmly adheres; as alſo near the *Scapula*, *Serratus Major Anticus* will ariſe with him, without mature and deliberate obſervation hereof: this brings the Arm backwards, ſometimes upwards; its diverſity of Fibres contracting themſelves doth occaſion theſe variety of Motions.

This is ſhewn at *Tab.* 13. at *C.C. D C.C.* Shews its Origination, *D.* Shews where its Tendon ariſeth, *N. N. O. O.* Shews the ſame laid bare at *Tab.* 15.

This brings the arm backwards to the Fundament.

Vſ.

Rhomboides.

This brings the Scapula backwards.

THis is fo called from *Rhombus*, the which is by Mathematicians faid to be a fourfquare Figure, not confifting of equal Angles, but of Lines: it arifeth broad and Flefhy immediately under the *Cucullaris*, from the Spinal proceffes of the three lower Vertebres of the Neck, and three upper Vertebres of the *Thorax*, whence running with thin Fibres, it defcends down to the *Scapula*, to whofe whole Bafis externally, it adheres broad and Flefhy; in the raifing of this, take care that you do not raife the Tendon of *Serratus Pofticus Superior*, who lies juft under him. And alfo becaufe it taketh its Origine from the three lower Vertebres of the Neck, and the three upper of the *Thorax*, and is inferted into the Bafis of the *Scapula*, whereupon it doth affift the *Cucullaris*, and bind the *Scapula* backward to the lower part of the Neck and upper part of the *Thorax*.

Ufe.

This Mufcle holds the *Scapula* to the Back: hence Confumptive people have prominent *Scapula's* from the weaknefs of this Mufcle; from this alfo in fome Perfons may an account be given of the Gibbofity of this part.

This you have at *Tab.* 15. at *B. B. in fitu, c. c. c. c.* Shews its Origination, *G.* Shews the fame laid bare at *Tab.* 24. *h. h. i.* Shews its beginning and ending.

Levator

The Explanation of the Fourteenth Table.

A. A. Cucullaris.
B. B. Shews one part thereof laid bare.
C. C. Levator Patientiæ.
D. D. Rhomboides.
G. G. G. G. Latissimus Dorsi.

Levator

fol.91

TAB. XIIII.

Levator Patientiæ.

THis is also called *Scapulam Attollens*: This ariseth from *This brings it forwards.* the second, third, fourth, and fifth Transverse Processes of the Neck, and hath as many Originations with their *Interstitia*: these joyning do descend, and adjoyn to the whole upper side of the *Rhomboides*, until it is inserted by a broad, Fleshy Tendon to the highest, as also to the lower Angle of the *Scapula*, and doth draw the same upwards and forwards, and is raised with the Arm. *Use.*

This Muscle had its name from *Spigelius*: it bearing many heavy burdens, for the *Scapula* by the help hereof with the Arm is lifted up forwards, and hence takes the better name of *Patientiæ*.

This you have at *Tab.* 15. at *A.* laid bare, 1. 2. 3. 4. Shews its four beginnings.

Rotundus

Rotundus Major.

This draws the Arm down backwards.

THis is also called *Humerum Deprimens*, its also called *Rotundus*, it being Sphærical; it ariseth with a Fleshy beginning from the lower Angle of the *Scapula*, and adheres Fleshy from his lower Rib, and having marched half way, deserts, and being indifferently dilated somewhat upwards, is inserted by a short, broad, and strong Tendon into that part of the *Os Humeri* where *Pectoralis* hath his insertion, *Use.* and doth draw the Arm down backwards. It partly lies under the Arm-pit.

Obs. Its generally observed by all Masters of Anatomy, that all Depressers are much less then the Attollers in Human Body. Thus the *Temporalis* is the strongest Attollent of the lower Mandible, and *Biventer* is the most infirm Depriment, and the reason is, there is required much more for the lifting up of a weight, than for his bringing down.

This you have at *Tab.* 15. at *E*.

Superscapularis Superior five Supraspinatus.

THis is Fleshy and long, and presseth the whole *Scapula* upwards: it ariseth from the whole Basis of the *Scapula* Fleshy above the Spine, filling the whole Cavity between the Spine and the upper Rib of the *Scapula*, and going back to the neck of it, passeth under the second Ligament of the *Humerus*, as the Biceps doth, and is inserted by a broad and strong Tendon obliquely into the neck of the *Os Humeri*, and doth bring the Arm about with the former; others do affirm that it moves the Arm upwards with the *Deltois*. And I apprehend that a great Use of this Muscle is to help the *Coracobrachialis* and *Infraspinatus*, in lifting up the Arm.

This you have at *Tab.* 15. at *F. L.* Shews the same laid bare.

This brings it about outwards.

Use.

Superscapularis Inferior sive Infraspinatus.

This brings the Arm about outwards.

THis covereth the whole Exteriour part of the *Scapula*, which under the Spine, arising from almost the whole Basis of the lower part of the *Scapula*, and possessing the *major* part of that Cavity, Fleshy, running backward, narrows himself according to the form of the part, and by a broad Tendon is inserted into the Ligament of the *Os Humeri*, as some will have it; But I humbly conceive, That this Muscle according to its situation, doth more probably assist the *Deltoeides* and *Coracobrachialis*, lifting the *Os Humeri* upward.

Use.

This you have at *Tab.* 15. at *G. M.* Shews the same laid bare.

Nonus

Nonus Humeri Placentini sive Rotundus Minor.

Fallopius calls this *Transversalis Brevior* from its *site*, and *Rotundus* from its form: It ariseth sharp and Fleshy from the lowest Angle of the *Scapula*, at his Basis, and growing more Fleshy to its Venter, decreases again, and terminates himself by a sharp Tendon into the neck of *Os Humeri*; This by some Anatomists is held as a part of that Muscle called *Rotundus Major*. [This helps the motion of Rotundus Major.]

This Muscle depresseth the *Os Humeri*, and is an Antagonist to the *Deltoeides* and *Coracobrachialis*. [Use.]

This you have at *Tab.* 17. at *Fig.* 2. at *A.B.C. A.* Shewing its Fleshy beginning, *B.* Its Cavity or Fissure, *C.* Its thin and Nervous Tendon; This you have also at *Fig.* 1. *Tab. id.* at *P. P. Q. Q. Q.* Shews the Nerve that passes through this into other Muscles.

Subscapularis.

Subſcapularis.

This brings the Arm inwards.

THis is ſeated in the Cavity of the *Scapula*, and poſſeſſeth the whole Cavity thereof: it ariſeth Fleſhy from the whole inward Baſis of the *Scapula*, and ſo running forwards, according to the dimenſions of the Bone, narrowing himſelf, and by a broad Tendon, is inſerted into the third Ligament of the *Os Humeri*.

Uſe.

Theſe laſt Muſcles do work the whole Arm about, from whence they have alſo their names, but the *Subſcapularis* brings it inwards, the *Superſcapularis Inferior* outwards, and alſo upwards, but theſe together do moderately elevate it; their Uſe may be moſt commodiouſly ſhewn, if the whole Arm be put into a middle Figure, and afterwards, neither abduced from the Breaſt to the Arm, neither the middle Figure changed with the Cubite of the Arm, the Cubitus being drawn outwards, and again brought inwards.

This you have at *Tab.* 11. at *C. C.*

The Explanation of the Fifteenth Table.

A. *Shews* Levator Patientiæ.
B. B. Rhomboides.
C. C. C. C. *Shews the Spines of the Vertebres, from whence they do take their Origination.*
E. Rotundus.
F. Superscapularis Superior.
L. *Shews it laid bare.*
G. Superscapularis Inferior.
M. *Shews it laid bare.*
a. Octavus Humeri Placentini.
v. *Shews it laid bare.*

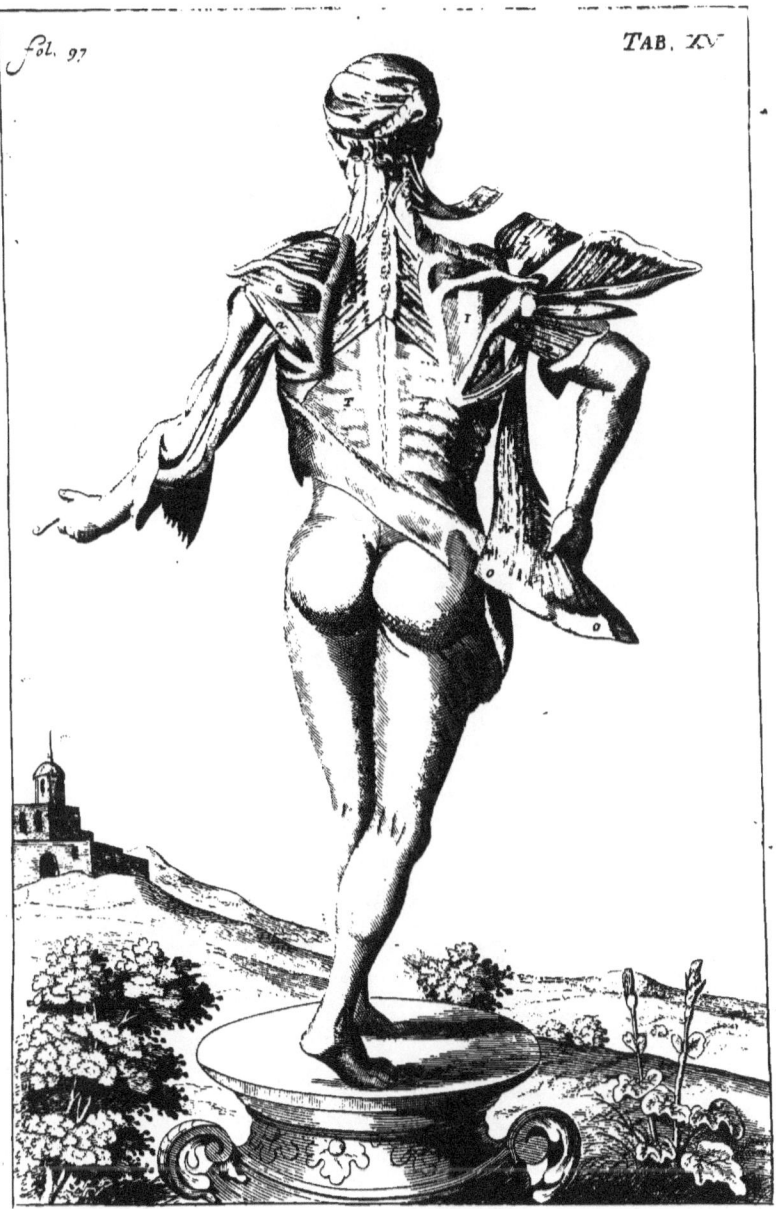

If the whole Arm with the Scapula *be taken off, the diſſection of theſe following Muſcles will with more eaſe be performed.*

Deltois.

THis is accounted the ſecond Muſcle of the Arm, called *This lifts up the Arm.* by ſome *Triangularis Humeralis*, ariſing generally Nervous; Firſt, forward from the middle of the *Clavicle* where it is broadeſt, next the *Os Humeri*; Secondly, from the top of the *Scapula*, where it is joyned to the *Clavicle*; Thirdly, from the whole Spine of the *Scapula*, theſe Originations preſently growing Fleſhy, deſcending and narrowing become a ſtrong Tendon, externally Fleſhy, internally Nervous, which is tranſverſely planted under the neck of *Os Humeri*; this *Uſe.* lifts up the Arm ſometimes directly, ſometimes forwards, or backwards according to its Series of Fibres, theſe being contracted: In the middle part hereof unexperienced Chirurgeons *Caution.* do make *Fontanels*, but very inconſiderately, becauſe this Muſcle being contracted, the Orifice of the *Fontanel* therein made is alſo therewith contracted, ſending thereby the Pea forth with force and pain, and the Iſſue doth heal up preſently; which Errour may with eaſe be ſhunned, if they made them in the diſtance between this and the *Biceps*, four or five Fingers breadth from the joynt of the Arm, in which place, when the Arm is bent, there is preſently perceived this Interſtitium.

This you have at *F. F. Tab.* 16. *H. H. I.* Shews the ſame laid bare.

Bb Biceps.

The Explanation of the Sixteenth Table.

A Serratus Major Anticus.
  æ. æ. æ. Shews its Originations.
B. Pectoralis. *c. b d. Shews its diverſity of Fibres.*
F. F. F. F. Deltois. H. H. I. *Shews it laid bare.*
L. M. Biceps in ſitu.
K. *Shews the ſame ſomewhat laid bare.*
X. *Shews* Brachieus in ſitu.

Octavus

Fol: 98 TAB. XI.

Octavus Humeri Placentini, sive Perforatus seu Coracobrachialis.

THis ariseth by a short and Nervous Origination from the *This drawn* Coracoidal process of the *Scapula*, near the Origination *the Arm upward.* of the *Deltois*, and so descending, becomes somewhat long, large, and Fleshy, and is implanted by a strong Tendon into the middle of the *Os Humeri*, and doth adduce it with the *Pectoralis* forwards to the Breast, as some would have it; the Venter hereof is perforated, and through the body of this Muscle doth pass a Nerve, which serves the rest of the Muscles; *Riolan.* reckoned this Muscle to be a part of the *Biceps*, or of the first Muscle of the Cubite, which doth contradict *Autopsy*.

And I do humbly conceive that this Muscle in truth doth assist the *Deltoides* in bending the Arm, and lifting up the *Os Humeri*.

This you have at *Tab.* 15. at *a.* in its place, *b.* Shewing the same laid bare.

Brachiæus

Brachiæus Internus.

This bends the Cubite forwards.

THis Muscle lodgeth under *Biceps*, but is more short and Fleshy: it ariseth about the middle of the *Os Humeri*, near the insertion of the *Deltois*, and *Pectoralis* double, broad, and Fleshy: partly by the Exteriour, partly by the inner side of the said Bone, but presently joyning, and firmly adhering to the Bone, in its descent becoming large, till at last it terminates broad and Fleshy in the Ligament of the Joynt, as well as in the Appendices of the Cubite and *Radius*; This with the *Biceps* doth very strongly contract the Cubite, lifting it up, being an assistant to the *Biceps*

Use.

This you have at *M. N. O.* at *Tab.* 17. *M.* Shewing its beginning, *N.* Its Venter or Belly, *O.* Its large and Fleshy Tendon.

Gemellus

Gemellus Major, Brachiæus Externus, aut Longus.

THis Muscle ariseth doubly broad and strong; partly Fleshy, partly Nervous, from the lower part of the *Scapula*, where it hath a peculiar Cavity a little under the neck of it: and growing Fleshy descends, and joyns himself to the inner side of the *Os Humeri*, where he meets with his other Fleshy Origination, and so making one, is carried down to the inner side of the *Process* of the *Olecranum*, and is there implanted; The Use of this Muscle is generally reputed to extend the Cubite, and is an Antagonist Muscle to the *Biceps Internus*, which is a Flexor and lifter up of the Cubite: and the *Biceps Externus* doth extend it, and put it backward and depress it. *This extends the Cubite. Use.*

This and its following partner you have described at R. R. S. S. Tab. 17. R. R. Shewing its beginnings, S. S. Its double Venter.

Extensor Cubiti Breuis, vel Brachiæus Internus.

This doth assist the former in its extension.

THis is the second of the Extenders of the Cubite, and doth arise backwards Nervous from the neck of the *Os Humeri*, and so becoming Fleshy, descends to the middle of *Os Humeri*, and doth inseparably mix himself with the former, and at length is inserted partly Fleshy, partly Nervous into the outward side of the *Olecranum*, about the place we lean on. These two Muscles as they have right Fibres given them from their beginning to their ends, they do extend the Cubite directly, as the two former did contract it.

Use.

Veslingius says this Muscle arises from the middle of the *Os Humeri*, and it is accordingly represented so in his Cuts.

This and its partner you have shewn at R. ℞. S. S. at *Tab.* 17.

Anconæus.

The Explanation of the Seventeenth Table.

M *N. O.* Brachieus Internus.
R. R. S. S. Gemellus Major.
Gemellus Minor *you have alſo in the ſame Table.*

Anconæus.

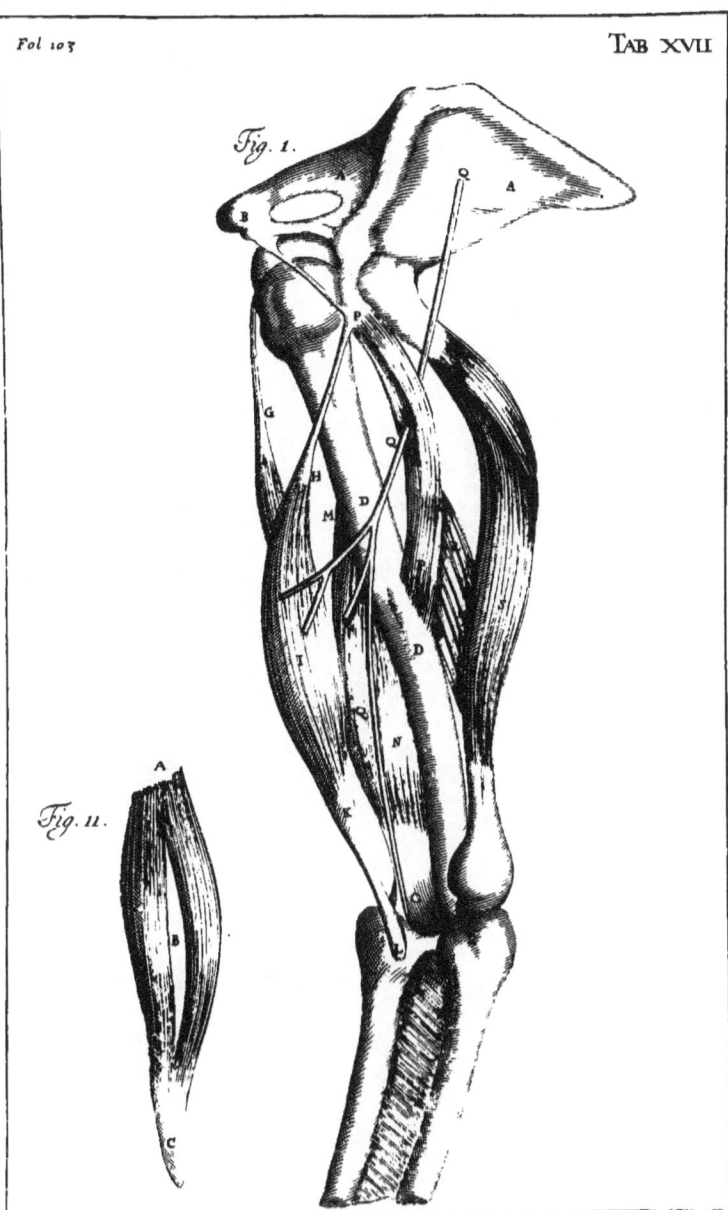

Anconæus.

THis being but a small bodied Muscle, doth arise from the lower and back part of the *Os Humeri*, being planted between the *Cubitus* and the *Radius*, and is inserted with a Nervous Tendon (sometimes obliquely) into the lateral part of the *Ulna*, a Thumbs length below the *Olecranum* or Elbow; As those two last do help forward the Cubites extention, this doth help the former, *&c.* *This doth also extend the Cubite. Use.*

This is not to be shewn or represented by Figure.

Palmaris.

(104)

Palmaris.

This moves the Palm of the Hand.

This ariseth round and Nervous, from the inward Extuberance of *Os Humeri*, and presently becoming Fleshy and narrow, about the middle of the Cubite is carried somewhat obliquely, and is turned into a long and round Tendon, and passing over the inward Ligament of the *Radius* comes to the Palm and there doth expand it self into a most broad Tendon, and is laterally inserted into the first Joynts of the Fingers, and is so closely fixed to the *Cutis*, that it is not thence to be separated without difficulty;

Use.

The contraction of this *Cutis* occasioneth apprehension, and doth endue the Palm of the Hand with an Exquisite Sense, and the Skin becoming immovable by the gripe, it occasioneth a more steady keeping of any thing therein contained

obs.

The learned *Fallopius* hath observed this Muscle sometimes to be double in either Arm, and arising from the same place, that one doth end in a broad Tendon, such as already is described, the other, in the Transverse Ligament of the *Carpus*.

This you have at *Tab.* 18. *Fig.* 1. at *D. E.* Shews where it passeth into a Tendon, *b.* Shews the expansion of the same Tendon, *G. G. G. G.* Its Terminations; At the same Letter you have the same expressed at the second Figure *Tab. ejusd.*

Caro

Caro Musculosa Quadrata.

Near adjacent to the former, is planted a Fleshy Substance *This doth assist the former.* arising from the *Membrana Carnosa* under *Mons Lunæ*, where the eighth Bone of the *Carpus* is placed, and then marcheth under *Palmaris*, to the middle of the Palm, and is inserted into the outside of that Tendon that separates the little Finger from the rest; this makes the Hand hollow, and the *Use* hereof is perceived and declared in large contractions of the Palm, drawing the *Mons Lunæ* to the middle of the Hand: this is when we would make it appear very hollow.

This you have at *P. P.* at *Fig.* 2. *Tab.* 18.

Flexor Carpi Interior five Ulnaris.

This sends the Wrist.

THis ariseth sharp, Fleshy and Nervous from the inner *Apophysis* of the *Os Humeri*, and so running Fleshy the whole length of the Cubite, to which it adheres, hath its Insertion at the Wrist into its fourth Bone, partly Nervous, partly Fleshy, this passeth not under the Transverse Ligament, but is only wrapt up with the common Membrane of all the Muscles; Its Use is thus, such as are the Inflectors do possess the whole part, and do arise from the inward Protuberance of the Arm: those that do extend, do arise from the back part and outward Extuberance.

Use.

This you have at *Tab.* 18. at *Fig.* 1. at R. R. *f.* Shewing its Fleshy and Nervous beginning, *g.* Its end partly Fleshy, partly Nervous; This also you have at the second Figure at the same Table, at the same Letter: K. K. Shews the same *in situ* at *Tab.* 19. at *Fig.* 1. *b.* Shewing its beginning, *c. d.* Its end as formerly.

Flexor Carpi Exterior five Radialis.

THis ariseth as the former from the same Extuberance, and so running somewhat transversly near the outer part of the *Primi Digitorum Flexores*, is fixed to the *Radius*, and a little before it arrives at the *Carpus* doth become a round Tendon, which cleaving to the Transverse Ligament, runneth under him, and enlarging himself, is inserted into that *Os Metacarpi* which stands before the little Finger; these two do contract the Hand. *This doth help the former in its Contraction.*

These two working together, the *Carpus*, and the Hand with it is also contracted; one only working, its sometimes moved into that side somewhat obliquely which is contracted. *Use.*

This you have at *Tab.* 18. *Fig.* 1. & 2. at S. S. *in situ*, L. Shewing its beginning, M. Its slender and Nervous Termination, L. L. Shews the same at *Tab.* 19. *in situ*, c. Shews there its Fleshy beginning, f. The Tendon of this Muscle, G. Shews the same laid bare at *Tab.* 20. *Fig.* 1.

Flexor

Flexor Secundi Internodii Perforatus.

This contracts the second Joynt of the Fingers.

THe Fingers, which are the great Messengers of Writing our Minds, and which are implanted in us for performing of many strong and vigorous Motions, do act these by the benefit of Muscles, by which they are both contracted, extended, and brought to a lateral Use; We begin with the Contractors, these being seated in the Cubite, and those generally do gain the name of Contractors which do bring the four Fingers into a Curvation; Of these Muscles of the Fingers, this is said to be the first, and taketh its name from its *Use*. Use, contracting the second Bone of the Fingers.

It ariseth from the inward Protuberance of *Os Humeri*, under the former, and so growing broad and Fleshy about the middle of the *Cubitus* and *Radius*, somewhat adhereing thereto marcheth forwards, and becomes wholly round and Fleshy near the Wrist, where it is divided into four Fleshy Portions, from which proceeds so many Tendons, all which are involved in one proper Mucaginous and thin Coat, and so running internally under the Ligament, gets through the Palm, and then doth distribute to the first and second Joynts of the four Fingers, as many Tendons which are perforated a little before their insertion for the transmission of the Tendons of *Tertii Internodii Flexor* This Muscle by some is called *Sublimis & Perforatus*.

This you have at *Q. Q. Tab.* 18. at both Figures *in situ, d.* Shews its Origination, *e. e.* Its Bivaricated Tendon, *D. D. D. D.* Shews the same at *Tab.* 19. *Fig.* 1. *C.C.C.C.* Shews it laid bare at *Fig.* 2.

Flexor

The Explanation of the Eighteenth Table.

D Palmaris *at Fig.* 1. *E. Shews where it paſſeth into a Tendon.*
F. *Declares where it expands it ſelf into a broad Tendon.*
G G. G. G. *Shews the Termination of theſe Tendons.*
0. 0. 0. 0. Flectentes Pollicem.
P. Caro Muſculoſa Quadrata.
O. Q. Flexor Secundi Internodii Digitorum.
⸺ing its beginning at Fig. 2. e. e. *Its Bivaricated Tendons.*
  Flexor Carpi Interior, *f. g. Shews its Origination and Ter-*
  nation.
  Flexor Carpi Extèrior, L. M. *Shews its Origination and Ter-*
  mination
V. *Extenſor Carpi Exterior.*
X. X. Pronator Radii Teres.
Y. Y. Tertium Pollicis Internodii Flectens.
Z. Z. Pars quædam Flexoris Tertii Internodii Digitorum.

Flexor

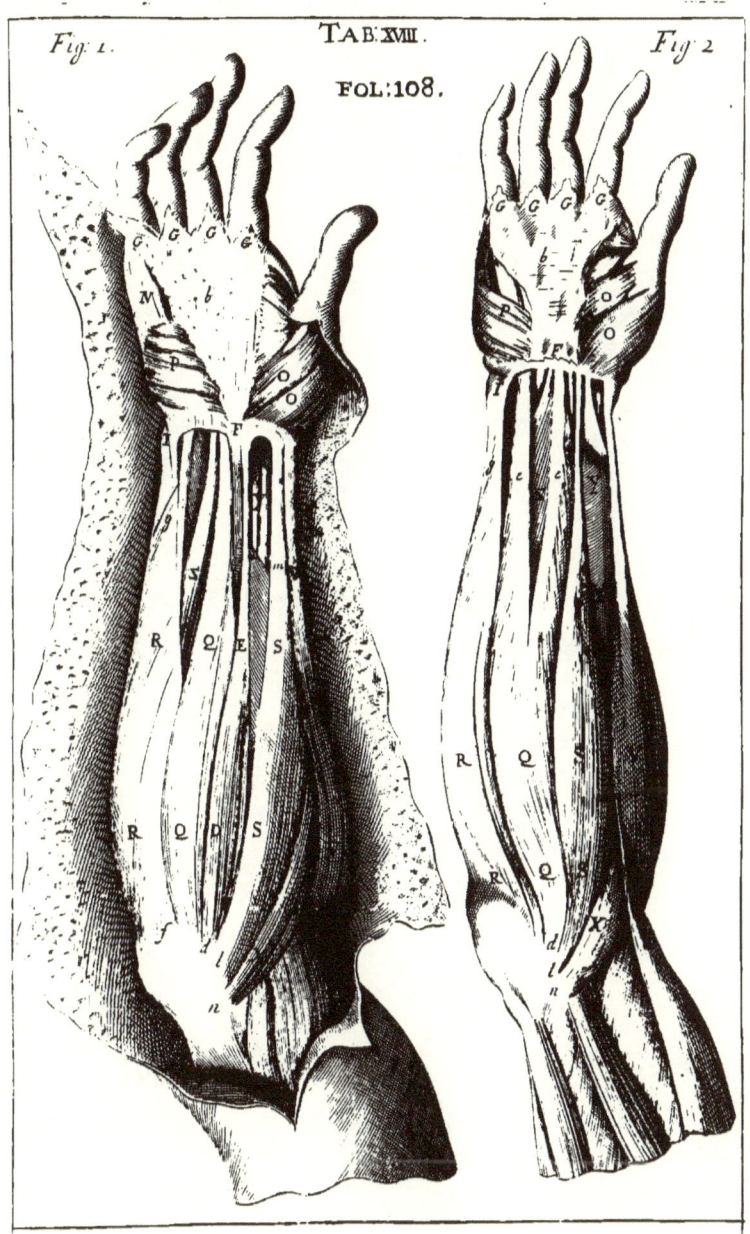

Flexor Tertii Internodii.

This doth contract the third Joynt, and from thence doth *This contracts the third Joint of the Finger.* takes its name: by some it is called *Profundus*, as also *Perforans*; it ariseth Membranous some part of him, from the same Extuberance as the former, the rest from the Root of the forward process of the *Olecranum*, becoming afterwards exactly round and Fleshy, running under the former, and adhering inwards to the *Cubitæus*, on the outward to the *Primi pollicis Flexor*, and descending about the middle of the Cubite, they divide, and make their progress through the Fissures of the others, and are inserted into the third Joynts of the Fingers: and that a right inflection may be made by these Muscles, and that the contracted Tendons may not rise up, and so lift up the *Cutis* in the inward part of the hand according to their length, a Channel being made out of the hard Membranes, they are inwardly included with a fatty and Oleaginous Humour in which they have their free course of Motion.

These Muscles are allowed to contract the third Joynts of the Fingers. *Use.*

This you have at *Tab.* 18. at Z. *Fig.* 1. & 2. This you have exactly at *Tab.* 19. *Fig.* 2. at D. D. *a. a. a. a.* Shewing its four Tendons.

Flexor Secundi Internodii Pollicis.

This bends the second Joynt of the Thumb.

THis *Flexor* is not seated in the Hand, as the former but in the Fleshy part of the Cubite: it ariseth round and Fleshy from the *Os Cubiti*, and so marching along by the *Radius* to which it adheres, as also to the Membrane that joyns the *Cubitus* and *Radius*, it comes to the Wrist, where it becomes a round Nervous Tendon, having also a proper thin Mucaginous Membrane, and so marching forwards, is inserted into the second joynt of the Thumb by a somewhat broad Tendon.

Use.   This contracts the second Joynt of the Thumb, to which it is fixed.

This you have at *Y. Y. Tab.* 18. *Fig.* 1. This also you have at *P. P. Fig.* 1. *Tab.* 19. This you have laid bare at *Tab.* 20. at *M. M.*

Pronator

Pronator Radii Teres.

The *Radius* is wrought with two Motions, the one *per accidens*, the other *per se*, and hence hath it given it two kinds of Muscles: the first from their Uses are called *Pronatores*, the other *Supinatores*. The second of the *Pronators* is called *Teres*, so called from its form, it ariseth from the Root of the inner prominence of the *Os Humeri*, and from the inside of the *Os Cubiti*, and is there joyned by a large Fleshy Origination to the *Radius*, and thence descending obliquely downwards by his inner side a little above the middle, is implanted into him Fleshy, from whence a Nervous Head doth proceed, which is inserted into the External Head of the *Radius*.

This brings the Wrist downwards.

This Muscle is held to bring the *Radius* downwards. *Use*

This you have at *X*. in both Figures *Tab*. 18. *n*. Shewing its Origination; This also you have at *Tab*. 19. *Fi*. 1. At *0 G. Fig*. 2. *ejusd. Tab.* you have the same at *K*. *d.* Shewing its Origination, *e*. Its Termination, *C.C.* Shews the same at *Tab*. 20. *Fig*. 1. *a*. Demonstrating its beginning, *b*. Its Termination.

Pronator

Pronator Quadratus five Inferior.

This works as the former.

THis is allowed the firſt of the *Pronators*, it is wholly Fleſhy, and ariſeth from the lower and inner part of the Cubite near the *Radius*, running over the Ligament that joyns the Cubite and the *Radius* tranſverſly, and ſo doth implant himſelf Fleſhy into the inward part of the *Radius*, with a broad beginning, much repreſenting a Mathematical Square, having four equal ſides.

Uſe.

Theſe two Muſcles, while they are contracted towards their Originations, do move the *Radius* forwards, and ſo doth connect the Hand which the lower part of the *Radius* by the benefit of the *Carpus*.

This you have at *Tab.* 19. *Fig.* 1. at *I. K.* Shews this, *Tab.* 20. *Fig.* 1. *e.e.* Shews its beginning, *f.f.* Its Termination.

Flexores

Flexores Primi Internodii five Lumbricales.

These are small and slender Muscles, arising round, long, and slender, from those Membranes which do enwrap the Tendons of *Tertii Internodii Flexores*, and so passing on Fleshy, are inserted by a round, Nervous Tendon into the first of the Joynts of the Fingers, and are best shewn by raising one from his Origination, and leaving him in his Insertion, and the other *è contrario*, these bends the Fingers laterally. *These bend the Fingers laterally.*

Use.

This you have at *F. F. F. F. Tab.* 19. *Fig.* 1. *G. G. G. G.* Shews their four Tendons, *F. F. F. F.* Shews the same at the same *Tab. Fig.* 2. *b. b. b. b.* Shewing their Originations, *F. F. F. F.* Shews the same in its place, *Fig.* 21. *Tab.* 21. *f. f. f. f.* The beginning of these Muscles, *g. g. g. g.* Shews their Tendons.

F f   Flexor

Flexor Primus, Primi Internodii Pollicis.

This bends the Thumb.

THe Thumb is bent or contracted in its Joynts, and these after several ways, and hence have they their names severally bestowed upon them, some of these being Contractors of the first, others of the second.

This first ariseth Fleshy from the upper seat of the Annulary Ligament which is in the *Carpus*, near the Thumb, and ascending, doth encompass the first and second Joynts of the Thumb, and is inserted into the head of the first Joynt Fleshy.

This you have at *Tab.* 19. *Fig.* 1. at *P. P.* This you have also at *A. B. Tab.* 21. *Fig.* 1. This you have also at the second *Fig. ejusd Tab.* at *B.*

Secundus.

The Explanation of the Nineteenth Table.

F I G. I.

C *C. Palmaris laid bare.*
D. D. D. D. Flexor Secundi Internodii Digitorum. *E. E.*
E. E. Shews its four Tendons.
F. F. F. F. Lumbricales. *G. G. G. G. Shews these four Tendons.*
H. Caro Musculosa Quadrata.
K. K. Flexor Carpi Interior.
L. L. Flexor Carpi Exterior.
O. G. Pronator Radii Teres.
P. P. Pollicem Flectentes.
Q. Minimum Digitum Abducens.
R. Pollicem Abducens.
S. S. Carpum Extendens Exterior.

F I G. II.

C. C. C. Flexor Secundi Internodii Digitorum *laid bare.*
D. D. Flexor Tertii Internodii. *a. a. a. a. Shews its four Tendons.*
F. F. F. F. Lumbricales, *b. b. b. b. Shews their Originations.*
G. G. Flexor Tertii Internodii Pollicis *laid bare. c. Shews its Tendon.*
I. Quadratus in situ.
K. Secundus Radii Teres. *d. Shews its Origination.*
O. Primus Secundi Pollicis Articuli Flexor.
P. Flexor Primus Primi Pollicis Internodii.
Q. Pollicem Abducens.

TAB. XIX fol. 115

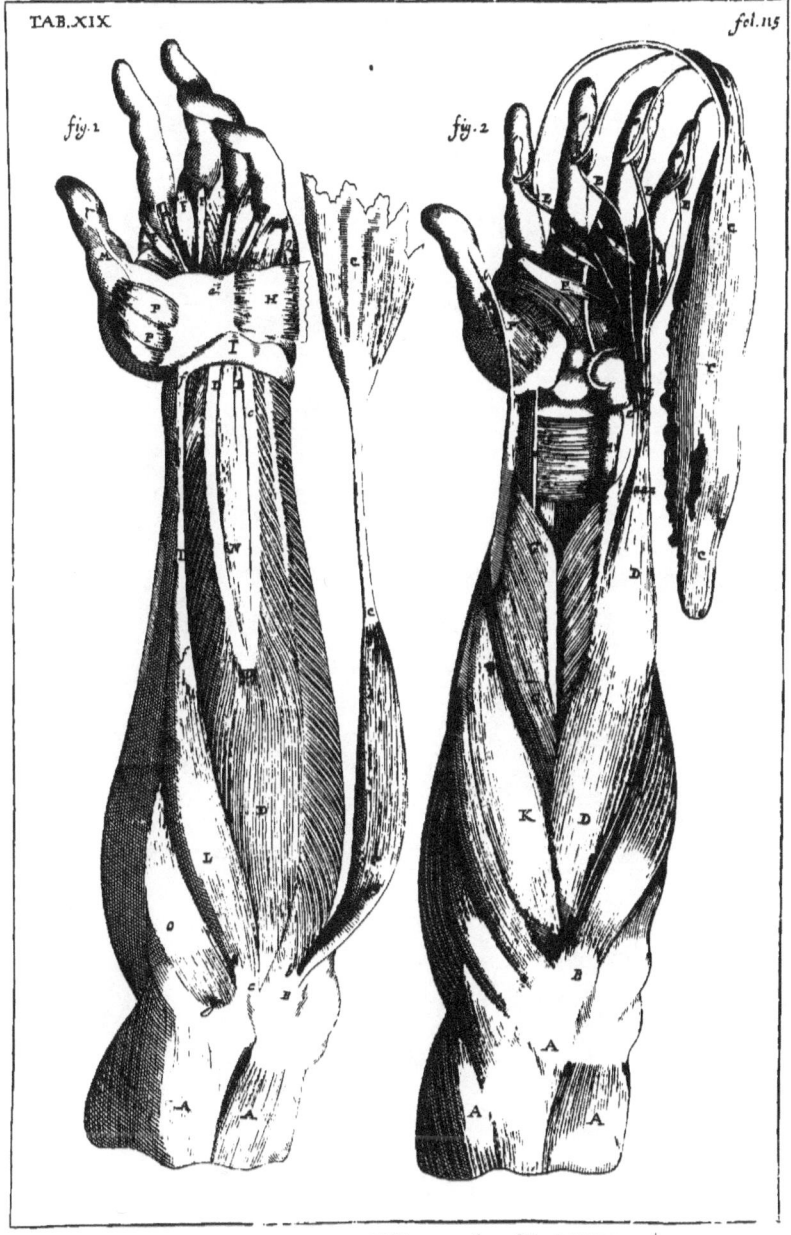

Secundus.

THis being smaller then the former, ariseth Fleshy, partly from the same Ligament, and partly from *Os Carpi*, next the Thumb, and running under the other, is implanted into the same Joynt from its Root, even to its middle: this is wholly lodged under the former, and is covered over with its breadth. *This helps the former in its contraction.*

These two do bend the Joynt of the Thumb, and adduceth *Use.* it to the Hand. *Riolanus* doth not acknowledge this as a *Flexor*, but rather doth think, that the Muscles arising from the Bones of the *Carpus*, and *Metacarpus*, to be either *Adductors* or *Abductors*.

This you have at *Tab.* 19. at *M. M. Fig.* 1.

Secundi Internodii Pollicis, Flexor Primus.

This bends the second Joynt of the Thumb.

This Muscle ariseth broad, thin, and Fleshy, from that *Os Metacarpi* that receiveth the Fore-finger a little below his head, and running towards his Thumb grows somewhat triangular, and is inserted by a Membranous Tendon into the head of the second Joynt of the Thumb on the side next the Fore-finger.

Use. This Muscle by most Anatomists is allowed to bend the second Joynt of the Thumb.

This you have at *0. 0. 0. 0. Fig.* 1. & 2. *Tab.* 18. *C.* Shews the same laid bare at *Tab.* 21. *Fig.* 2. This is also shewn *in situ* at *Tab.* 18. at *i. Fig.* both.

Secundus.

Secundus.

THis ariseth with a broad and Fleshy beginning, from *This works as the former.* the middle part of *Os Metacarpi* of the third Finger, running to the Thumb, and is inserted into the middle of the second Joynt thereof, in its inward part.

This Muscle is said to be next to the former, and thus each do succeed one another in place and order; and as touching their proper Uses, you will meet them all summed up together in *Quarto Musculorum Digitorum*.

This you have at *Tab.* 18. at *O. O. I.*

Tertius.

Tertius.

This doth operate as the former.

This follows the second, and is contiguous to it, this ariseth broad and Fleshy from the *Os Metacarpi* of the the third Finger, and is implanted into the same with the former.

This Muscle is also allowed to begin where the other ended: all these together do make up that Fleshy Mass which our Chiromancers make use of, which you have more fully in the next; they are contracted according to the Bone moving towards the other Fingers, and according to their variety of Operations they do express their diversity of Uses.

This you have at *O*. 3. in *Tab.* 18. at both Figures; This you have laid bare at *D.D. Tab.* 21. *Fig.* 2.

Quartus.

Quartus.

THis ariseth Fleshy from the *Os Metacarpi* of the Little Finger about the middle of it, and running under the other, is inserted into the same Joynt as the former. *This brings the Thumb to the Little Finger.*

All these Muscles are Fleshy, of which, the two contracting the first Joynt, with that abducing the Thumb, do constitute that little Hillock of the Thumb so called by Chiromancers; but the third of the second Joynt maketh whatsoever is Fleshy between the Life-Line, and the aforesaid little Hill; these are contracted according as the rest of the Joynts of the Fingers, and they working together, do bring the Thumb to the side of the Little Finger, hereby working the Hand into a hollowness. *Use.*

This Muscle brings the Thumb up to the great or Fore-Finger.

This Muscle running under the former is not to be shewn by Figure.

The

The Explanation of the Twentieth Table.

FIG. I.

C C. Secundus Radii Terei.
 a. Shews its beginning, b. Its Termination.
D. D. Carpi Flexor & Extensor Internus *laid bare.*
G. Externus Carpi Flexor *laid bare.*
K. Quadratus.
e. e. *Shewing its beginning.*
f. f. *Its Termination.*
M M. Secundi Pollicis Internodii Flexores *laid bare.*
N. N. N. N. Interossei.
P. Minimum digitum Abducens.

FIG. II.

D. D. Primus digitorum Extensor.
a. Shews its beginning.
b. b. b. *Its threefold division,* f. f. f. *Shews three Tendons belonging to him.*
H. Pollicis Adductor.
L. M. Carpum Extendentes.

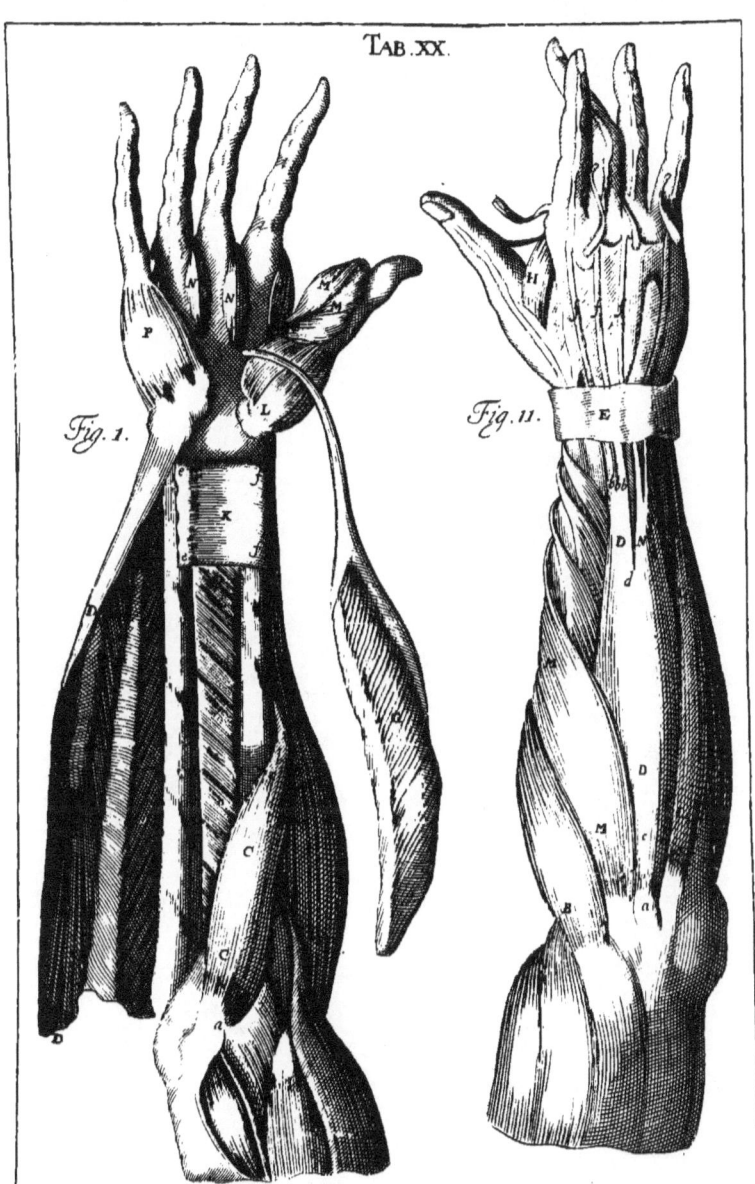

Minimi Digiti Abductor.

The Fingers besides Flexion and Extension, and being brought to the sides, are both adduced and abduced; then said to be adduced when they are drawn towards the Thumb; abduced, when they are retracted from thence: and the Muscles which do perform these Motions are many, some of which are accounted common, others proper; the common are commonly reckoned eight, the which from their *site* are generally called *Interossei*. *This abduceth the Little Finger.*

This Muscle called *Minimum Digitum Abducens*, is planted in the bottom of the Hand under the Little Finger, short and strong, arising Fleshy from the fourth Bone of the *Carpus*, and so extending it self by the *Metacarpe*, is inserted into the outward side of the first Joynt of the Little Finger.

This Muscle doth abduce the Little Finger from the rest of the Fingers, and hath given it this particular Use, that whilst we do apprehend Sphærical Figures, the same time, the Little Finger is abduced from the rest. *Use.*

This is call'd also *Hypothenar* by some Authors.
This you have shewn you at *Tab.* 19. at Q. *Fig.* 1. At *Tab.* 20. you have at P. in this place, At *Tab.* 21. *Fig.* 1. you have it *in situ*, And at *Fig.* 2. of the same Table it is laid bare at E. E.

Hh   Pollicis

Pollicis Abductor.

This abduceth it from the Little Finger.

THe Thumb when moved laterally, is either abduced from the reſt of the Fingers, or adduced; this *Abducens* ariſeth Nervous near the *Flexor Primi Internodii Pollicis*, from the Interiour part of the Bone ſuſtaining the Thumb; and then becoming Fleſhy, implants it ſelf by a Membranous Tendon into the firſt Joynt of the Thumb, and this abduceth it from the Little Finger.

Uſe.

This is call'd alſo *Thenar* according to the *Greek* Idiom which names the more protuberant parts of the Palm δύναμαι ἀπὸ τῦ δύνειν ἀ precutiendo.

This you have at *Tab* 19. *Fig.* 2. at *Q. & Fig.* 1. at *R. O.* Shews the ſame laid bare at *Fig.* 2. *Tab.* 22. Æ. Shews the ſame in place, *Tab.* 21. *Fig.* 1.

Pollicis

Pollicis Adductor.

Dducens is that which is seen in the space between the *Pollex* and the *Index*, and ariseth Fleshy from the outward and back part of that *Os Metacarpi* that sustains the Fore Finger, and is inserted Fleshy and broad to the inside of the Thumb to the first Joynt, and doth adduce the Thumb to the Index. *This brings the Thumb to the Little Finger.*

This is call'd *Antithenar* by some Authors.   *Use.*
This you have at *Tab.* 21. *Fig.* 2. at *H*. And at *O. Fig.* 2. *ejusd. Tab.* At *Tab.* 20. you have it at *Fig.* 2. at *H*.

Interossei.

Interossei.

These work the Fingers laterally.

THese *Interossei* are generally accounted eight, Fleshy and small, and long, arising from the Bones of the *Metacarp*, according to whose whole length they do march; when these do attain to the Roots of the Fingers, they become Tendons, and are laterally inserted from the first to the second Joynts; there are six of these planted in the three Joynts of the Bones of the *Metacarp*, (*viz.*) two in every one, so as one is carried to the inward, the other to the outward Finger, and another thereof belongs to the first Bone of the *Metacarp* which sustains the *Index*, and is incumbent in that part which respects the Thumb; the last adheres to the last Bone of the *Metacarpus* in the outward part of the Hand, or in its back thereof.

Use.

These Muscles by how much they do attain the rest, do also extend the second and third Bone, which is first observed by *Galen* 1 *de usu part.* 18. and this is the reason, why the Extensor of the Fingers being cut, yet their extension is not quite abolished: for these *Interossei* working together, do very excellently perform this Extension, and as they are implanted to the first Bone, if either of one of the Fingers be contracted, they do extend the first Joynt, if either, it either doth abduce or adduce it.

This you have at *f. G. H. H. H.* at *Tab.* 2. *Fig.* 2. At *Tab.* 2. you have the same at *N. N. N. N. Fig.* 1.

Extensor

Extensor Carpi exterior, sive Radiæus Externus, seu Bicornis.

THis is accounted the first of the outward Muscles of the *Carpus*, it ariseth from the External accuminated part of the Arm, with a broad and Nervous beginning, then growing more Fleshy, he runs by the *Radius* according to its length, and arriving half way, it marcheth into a strong Tendon, and is presently inserted with a double Tendon into the first and second Bone of the *Metacarpus*; and hence by reason of its double Insertion and Origination, is by some called *Geminus*. *This extends the Carpus.*

The Use of this Muscle is to extend the *Carpus*. *Use.*

This you have at *Tab.* 22. *Fig.* 1. at E. B. B. Shews the same, *id. Tab. Fig.* 2. *a.* Shews its beginning, *b. d.* Its Tendinous Termination, *I. K. K.* Shews the same laid bare, *Tab.* 23. *Fig.* 2. *I.* Shews its beginning, *K. K.* Its two Venters, *b. b.* Declares its two Tendons produced thence, *L. M.* Shews them also at *Tab.* 20. *Fig.* 2.

The Explanation of the One and twentieth Table.

FIG. I.

Æ Pollicem Abducens.
B. Primum Articulum Pollicis Flectentes.
C. C. Abducens Minimum Digitum.
D. *Part of the Tendon of the third Flexor of the Thumb.*
F. F. F. F. Lumbricales.
f. f. f. f. *Shews their Originations.*
g. g. g. g. *Their Tendons.*
G. G. G. G. *The four Tendons of the* Flexores Tertii Internodii.

FIG. II.

B. Primi Pollicis Internodium Flexor.
a. Primi Pollicis Internodium Abducens.
C. C. Primus Secundi Pollicis Internodii Flexor *laid bare.*
D. D. Secundus & Tertius Secundi Pollicis Flexores.
E. E. Duo minimi digiti Abductores.
F. G. G. H. H. Interossei.
b. b. b. *The Tendons thereof.*

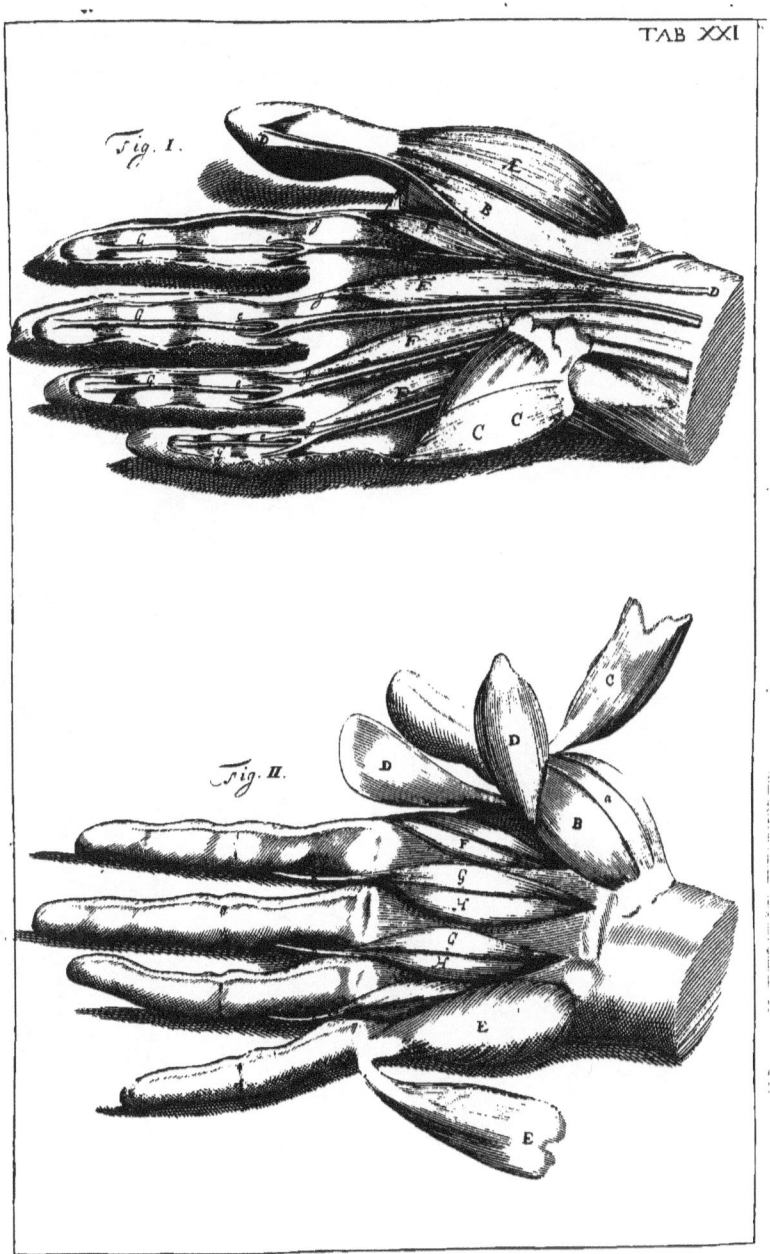

Extensor Carpi Interior, sive Cubitæus Internus.

THis ariseth from the Internal *Apophysis* of the Arm, as *This doth extend the Hand.* also from the top of the Cubite, and being dilated through the Cubite according to his length near the *Carpus*, is turned into a strong and round Tendon, by which he is inserted into a *Sinus*, above the lower Appendix of the Cubite, and into the upper part of the *Os Metacarpi* of the Little Fingers.

That worthy Observation of these parts is, that all the *obs.* Flexors are planted in the forepart, and do arise from the inward protuberance of the Arm, and that the Extensors are seated in the back part, and do take their Origination from the outward Extuberance: And thus the Hand by the benefit of the *Carpus*, joyned with the *Radius*, is made either to bend or extend it self.

This you have at *C. C. Tab.* 22. *Fig.* 2. *e.* Shewing its beginning, *f. f.* Its Tendinous end, *L. M.* Shews both the Extensors also at *Tab.* 20. *Fig.* 2. *H.* The same laid bare at *Tab.* 23. *Fig.* 2. *e.* Shewing its beginning, *f.* Shewing towards its Tendon.

Digitum

Digitorum Secundi & Tertii Internodii Tensor Communis.

These do extend the second and third Joynts of the Fingers.

AS there were some Muscles which did inflect the Fingers, and others designed for contracting the Thumb, so also are there diverse Muscles appointed for extention of the same; This is nominated as the first of the Extenders of the Fingers, it ariseth partly Fleshy, and partly Nervous from the outward *Apophysis* of the Arm, and becoming more Fleshy, descending between the two Extensors of the *Carpus*, and then narrowing doth divide it self into three round Tendons, which are included in a common thin Mucaginous Coat, and so are carried under the Annular Ligament, passing which, they divide themselves, and are inserted into the first, middle, and third Fingers, at the second and third Joynts.

Use. These Tendons do reach to the ends of the third Bone, and do lodge under the Nails, whence follows, that such sharp and girding pains do happen in these parts when any Injury falls upon them.

This you have at *D. D. Tab.* 22. *Fig* 1.

Minimi

Minimi Digiti Tenfor.

THis ariseth sharp and Nervous from the same part as the former, and for near half way is united to it, so as they seem all as one, but coming to the transverse Ligament, it becomes a round, large, and Nervous Tendon, which is implanted into the first, second, and third Joynts of the Little Finger. *This doth assist in extension.*

This you have described with the rest.

The Explanation of the Two and twentieth Table.

FIG. I.

C. C. Digitorum Extenfor Primus *laid bare.*
   a. a. Shews its Tendon
D. D. Extenfor Digitorum Secundus.
E. Carpus Exterior Extendentium.
F. Indicem Abducens.
G. Pollicis Extenfor Primus.
H. Pollicis Extenfor Secundus.
O. Supinator Radii Brevis.

FIG. II.

B. B. Carpus Extendentium Exterior.
C. C. Carpus Interior Extendentium.
D. Supinator Radii Brevis.
H. Pollicis Extenfor Primus *laid bare.*
I. Pollicis Extenfor Secundus *laid bare.*
O. Pollicem Adducens.

Supinator

Fol 129   TAB. XXII

FIG. I   FIG. II

Supinator Radii Longus.

This is called *Longus*, becaufe it obtaineth the longeft belly of all the Mufcles which do creep about the Cubite. *This brings the Radius outwards.*

This arifeth from the middle of the *Os Humeri*, and running obliquely over the *Radius*, at its bottom it becomes a Membranous Tendon, and is faftned to the upper part of the Appendix of the faid *Radius*, inclining fomewhat inwards.

If this and its Companion be contracted towards their Originations they do move the *Radius* forwards, as alfo the Hand, the which is tied to the lower part of the *Radius* by the benefit of the *Carpus*. *Ufe.*

This you have at *Tab.* 23. *Fig.* 1. at *C. a.* Shewing its beginning, *b.* Its Tendon, *D.* Shews the fame laid bare at *Fig.* 2. *Tab.* 23. *a.* Shews its beginning, *b. b.* Its Tendon.

Secundi

Secundi & Tertii Pollicis Tensor.

This extends the second and third Joynts of the Thumb.

THis ariseth from the same place of the Cubite as the former, and ascends obliquely over the *Radius*, and divides its self into two unequal parts, yet closely adhering, and is carried in a proper Channel at the Appendix of the *Radius*: the upper part remaining somewhat Fleshy, yet at last becomes a round Tendon, and is inserted into the *Os Carpi* which receives the Thumb; the other being presently subdivided into two small pieces of Flesh, do at length become Tendons: the first of which is inserted into the first Joynt of the Thumb, the other by a Membrane, fixeth its self to the second and third Joynts of the Thumb.

This you have at *Tab.* 22. at *H. & I. Fig.* 1. *l.* At *Fig.* 2. *ejusd. Tabul.* shews the same.

Indicem

Indicem Abducens.

THis ariseth with a Fleshy Origination from the middle of the Cubite, and so running obliquely to the Appendix of the *Radius*, it becomes two Tendons, which by a proper Sinus in the said Appendix are transmitted under the Annulary Ligament over the *Metacarp*, and the upper Tendon is carried to the Root of the Little Finger, the other is implanted into the Root of the second Finger obliquely, that it may abduce from the Thumb: it is accompanied with two Extensors of the Fingers in the lower Joynts, as the second and third. *This moves the Finger laterally.*

This you have at *F. Fig.* 1. *Tab.* 22.

L l   Supina

Supinator Radii Brevis.

This helps Longus in its Motion.

THis being shorter and thinner than *Supinator Radii Longus*, ariseth from the Exteriour part of the Ligament of the lower Head of *Os Humeri*, and from the process of the Cubite; and running obliquely (outwardly Membranous, inwardly Fleshy) doth recover the middle of the *Radius*, and is strongly implanted into it.

Use. These two Muscles if they do work together, they do contract the *Radius* forwards and outwards, and so do bring the Hand upwards: but the one draws the part downwards, and the other draws it upwards.

This you have at *O. Tab.* 22. *Fig.* 1. *D.* Shews the same at *Tab. ejusd. Fig.* 2. *D.* Shews the same at *Tab.* 23. *Fig.* 2. *c.* Shewing its Insertion into the *Radius*, *E.* Shews it at *Fig.* 2. *ejusd. Tab. c.* Shewing its Origination, *D.* Its Tendon.

Primi Internodii Extensores.

They arise Fleshy, round, long, and slender from the Bones of the *Metacarp*, and so running to the Roots of the Fingers, are laterally inserted to the first and second Joynts of the Fingers, and are to be raised as the *Flexores primi Internodii*; When these Muscles are contracted to their Originations, the second and third Joynts of the Fingers, together with the help of the *Interossei* are primarily extended, and in time of need, may serve to assist in oblique Motions: and wise Nature placed these inwardly, that if by any mischance the outward Tendons might receive any mischief, or happen to be wounded, yet by the assistance of these, the Fingers might be extended.

These do extend the first Joynts of the Fingers.

Use.

This you have at *D. D. Tab.* 20. *Fig.* 2. *a.* Shews its beginning, *b. b. b.* Its division into three Fleshy parts.

The Explanation of the Three and twentieth Table.

FIG. I.

D Supinator Radii Longus *laid bare.*
   a. Shewing its beginning.
b. b Its Tendon.
E. Supinator Radii Brevis.
C. Shews its Origination.
D. Its Insertion.
H. Shews the Ligament which distinguishes the External Muscles from the Internal.

FIG. II.

C. Supinator Radii Longus.
D. Supinator Radii Brevis, *c. Shews its Insertion.*
H. Carpum Extendens Interior *laid bare.*
e. Shews its begiming, f. Its Tendon.
I. K. K. Carpum Extendens Exterior *laid bare.*
b. b. Shews its two Tendons.

Fol.134.　　　　　　　　　　　　　　TAB.XVIII.

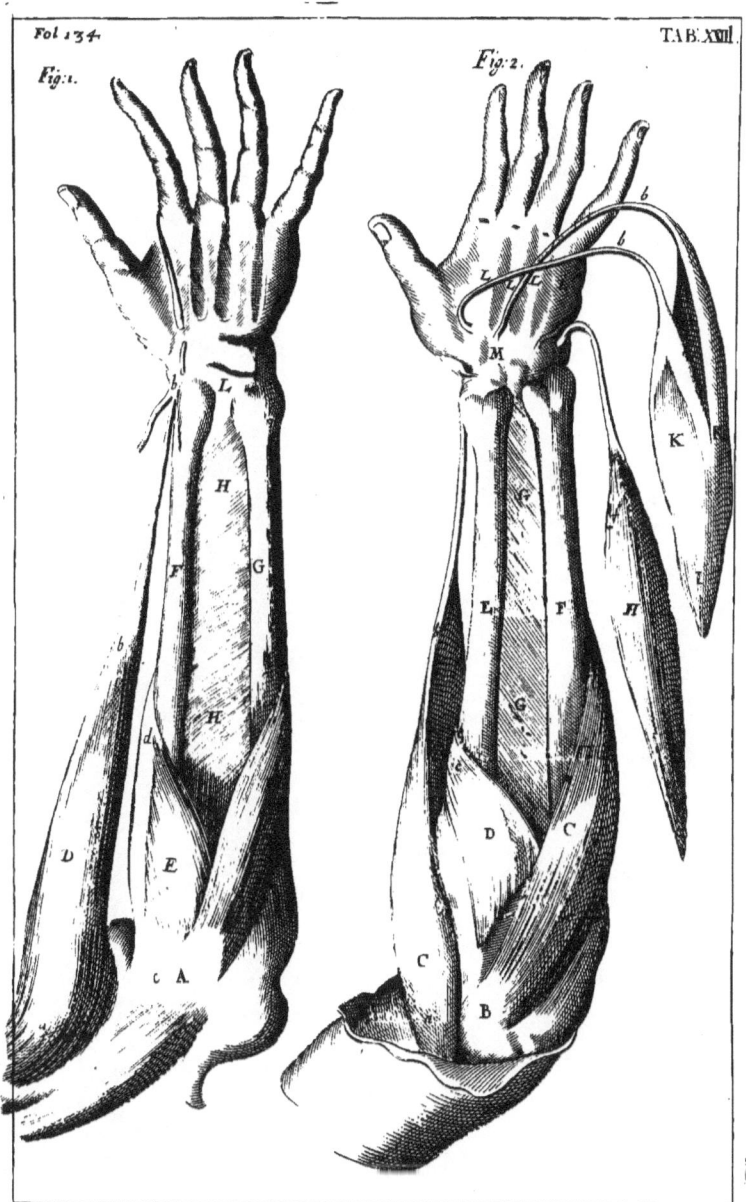

Fig:1.　　Fig:2.

Next come we to the Body it self as it lies.

Serratus Major Posticus.

TO the former Muscles of the *Thorax* are also added these two, so called *Serrati postici* from their Actions, as also from their Indentings: and *Postici* being contrary to those *Serrati* we have already described, and they are also called either *Superiores* or *Inferiores* from their places which they do possess. ^{These do dilate the Thorax.}

This Muscle being small is inserted in the Back under *Rhomboides*, between either *Scapula*, and upon the first pair of the Muscles of the Head, arising very thin and Membranous; from the Spinal processes of the three last Vertebres of the Neck, and the first of the *Thorax*, and in its oblique progress towards the Convex part of the *Thorax* it becomes Fleshy, and is inserted into the four upper Ribs, by so many distinct Terminations, and by drawing them upwards doth dilate the *Thorax*.

This you have at *Tab.* 24. at *C. in situ*, *d.d.* Shewing its beginning *e.e.e.* Its end consisting of three sorts of Fibres, *D.* The same laid bare.

Serratus Posticus Inferior.

This dilates the lower part of the Thorax.

THis Muscle is broad, thin, and Membranous, placed almost in the middle of the Back under *Latissimus*, and the *Aniscalptor* of the Arm, arising from the Spines of the lower Vertebres of the Back, and the first of the Loyns, and marching along transversly becomes Fleshy and is inserted into the four lower Ribs by so many distinct Terminations, the *Use.* which drawing outwards doth dilate the lower part of the *Thorax*.

E. Shews this at *Tab*. 24. *f.f.* Shews its Origination, *g.g.g.* Its Serrated Insertion, F. This Muscle laid bare.

Splenius

Splenius sive Triangularis.

This ariseth double: First, from the Spines of the fourth, *This brings the Head backwards.*
fifth, third, second, and first Vertebres of the *Thorax*: Secondly, from the Spines of the five lower Vertebres of the Neck, and so running broad and long about the third Vertebre of the Neck, both the Originations do unite: and by oblique Fibres both Muscles do insert themselves into the middle of the *Occiput*; You must take off from its Originations, and preserve as many of its *Ansulæ* as you can between the Spines by running between, and recovering its Tendon; If both move, they draw the Head directly backwards, if only one move, it turns the Head laterally.   *v;e.*

This you have at *Tab.* 24. at *A.A. A.A.* Shews the same at *Tab.* 26. This you have at *B. B.* also *Tab.* 25. *Fig.* 1. .

Trigemi

Trigeminus five Complexus.

This extends the Neck.

THis is the second pair of the Extenders, lying under the former, and is called *Trigeminus*, because it has allowed it a threefold Origination, and seemeth to be conflated out of three Muscles running into one: it hath various beginnings, and obtains both many and Nervous parts; it ariseth threefold, first, from the fourth and fifth transverse Processes of the Vertebres of the *Thorax* and immediately becoming Fleshy doth ascend over the rest of those Vertebres, until he reacheth the lowest Vertebre of the Neck, where it becomes a round Tendon; Not far from thence again it becomes Fleshy, and inserts it self into the middle of the *Occiput*; The second Origination is by a short round Nerve from the same Process of the last Vertebre of the Neck, and thence becoming Fleshy is joyned to the other before its Insertion; The third Origination is partly Fleshy, and Nervous from the transverse Processes of the first and second Vertebres of the *Thorax*, and running obliquely outwards, after union with the former, is inserted into the Root of the Mammillary Process, bestowing an Ansula upon every transverse Process of the Neck. To find this fairly, divide the sides of *Spinatus*, and *Longissimus Dorsi*, and his Origination will more plainly appear.

Obs. Riolanus doth observe that the Fibres both of this *Complexus* and *Splenius* to be intersected and disposed cross-ways for the better strengthning of either Muscle.

B. B. Shews this at *Tab.* 24. B. B. Shews the same at *Tab.* 26. C. Shews the same laid bare at the same Table.

Tranſverſalis.

THis ariſeth from the tranſverſe Proceſſes of the ſix upper Vertebres of the *Thorax*, and ſo growing thicker, is implanted externally into all the tranſverſe Proceſſes of the Neck, and hence had it given it its name, and doth draw the Neck backwards: but one of theſe only working, they bend it obliquely downwards; between theſe are carried the Nerves of the Spinal Marrow, paſſing out of the Vertebres of the Neck.

This extends the Neck.

Uſe.

This you have at *E. E. Fig. 2. Tab. 25.*

The Explanation of the Twenty fourth Table.

A. Triangulare *by some called* Splenius.
b.b. Shews its first and second sides.
B. B. Trigeminus.
C. Serratus Posticus Superior.
d.d Shews its Origination.
e.e.e. Shews its Termination.
D. *Shews the same laid bare.*
E. Serratus Posticus Inferior.
f.f. Shews its Origination.
g.g.g. Its Insertion.
F. *Shews the same Muscle laid bare.*
G. *Shews* Rhomboides *laid bare.*
h.h. Shews its Fleshy beginning.
i.i. Its Fleshy Termination.
H. H. Longissimus Dorsi.
I. *The same laid bare.*
K. K. Sacrolumbus.
L. M. N. *The same laid bare.*

Spinati

TAB XXIV.

Spinati Colli.

THis pair are long and large, possessing the whole Neck, between the Spines: it ariseth with many beginnings from the Roots of the Spines of the seven uppermost Vertebres of the *Thorax*, and ascending, gets a Tendon out of every transverse process of the Vertebres of the Neck, and is firmly implanted into the whole lower part of the second Spine of these Vertebres, and there the right and left do meet, and are so all the way united, that they are not divisible but by the Spine; These with the former do also extend the Neck, and then the Head, either directly, if they work together, or if they work singly or apart, it brings it obliquely.

This extends as the former.

Use.

F. f. f. These you shall have at *Fig.* 2. *Tab.* 25.

Recti

Recti Majores.

These extend the Head.

THese are the fourth pair: this Muscle is small, thin, and Fleshy, arising from the points of the Spines of the second Vertebre of the Neck, and ascending, are inserted into the middle of the *Occiput*, and doth help the motion of the third pair.

These are generally held to be the fourth pair of the Neck, and are granted to give assistance to the former in their Extentions.

This you have at *E. Fig.* 1. *Tab.* 26. At 25. you have them at *C. C.* At *b. b.* you have them, *id. Tab. Fig.* 2.

Recti Minores.

These pair lying under the former, being of the same substance and shape, accompanied with the like ductus; do arise from a small protuberance of the first Vertebre of the Neck round, and ascending, are implanted as the former underneath them; by the benefit of these *Majores* and *Minores*, if the whole pair work together, it extends the Head directly, but if one only move, it is moved laterally.

These do help the former.

Use.

Obs.

Nature hath made so many Muscles for extention of the Head, that Man might more aptly fit himself for the Contemplation of Cœlestial Bodies, and for this Motion, small Muscles were thought most requisite: and lest they might tire or grow weary in their long dependence, Providence hath ordered to these, long Muscles, more properly adapted for a longer bowing of the Head, or extending it to a sharper Angle.

These you have at *Fig.* 1. *Tab.* 25. at *D. D. Fig.* 2. *ejusd. Tab.* you have them at *b. b.*

O o   Obliqui

Obliqui Superiores.

These do turn the Head about.

THe sixth pair are planted under the *Recti* or the forementioned, answering their form or shape, being but small, and arising from the outward side of the *Recti* at their implantations, and obliquely descending, are inserted into the process of the first Vertebre of the Neck: if both move, they nod, and directly backwards: if only one, it inclines the Head laterally.

Use.

These are called *Obliqui* from their *site*, and one pair are implanted above another, either of which do lie under the *Recti Extendentes*, whose substance and form they exactly do answer. *Bauhinus* will have these to arise in the *Occiput*, and to end in the lateral processes of the first Vertebre of the Neck.

These you have at *F. Fig.* 1. *Tab.* 26. At *Tab.* 25. you have the same at *E. E. Fig.* 1. *Fig.* 2. *ejusd. Tabul.* you have them at *e. e.*

Obliqui

Obliqui Inferiores.

THis lower pair ariſeth longiſh, Fleſhy, and thin from the Spine of the ſecond Vertebre of the Neck, and obliquely aſcending, are inſerted with the *Obliqui Superiores* into the tranſverſe Proceſs of the firſt Vertebre. *This helps the former.*

When theſe are contracted, they work it about the Spine circularly, whence it is, that the Head conſiſting upon the ſame, it is alſo moved circularly therewith to the ſides; but becauſe theſe two pair are very ſmall, the two former pair of Extenſors, they do help thoſe in their Motion, as alſo thoſe of the right, as we have already ſhewn. *Uſe.*

Theſe you have at *Tab.* 25. *Fig.* 1. at *F*. And at *d.d.* you have them at *Fig.* 2. *ejuſd. Tab.*

The

The Explanation of the Twenty fifth Table.

FIG. I.

A Trigeminus.
B. Splenius.
C. C. Recti Majores.
D. D. Recti Minores.
E. E. Obliqui Superiores.
F. F. Obliqui Inferiores.

FIG. II.

a. a. Recti Minores.
b. b. Recti Majores.
c. c. Obliqui Superiores.
d. d. Obliqui Inferiores.
E. E. Transversales Colli.
f. f. f. f. Spinati Colli.

Longis-

TAB. XXV. fol.144.

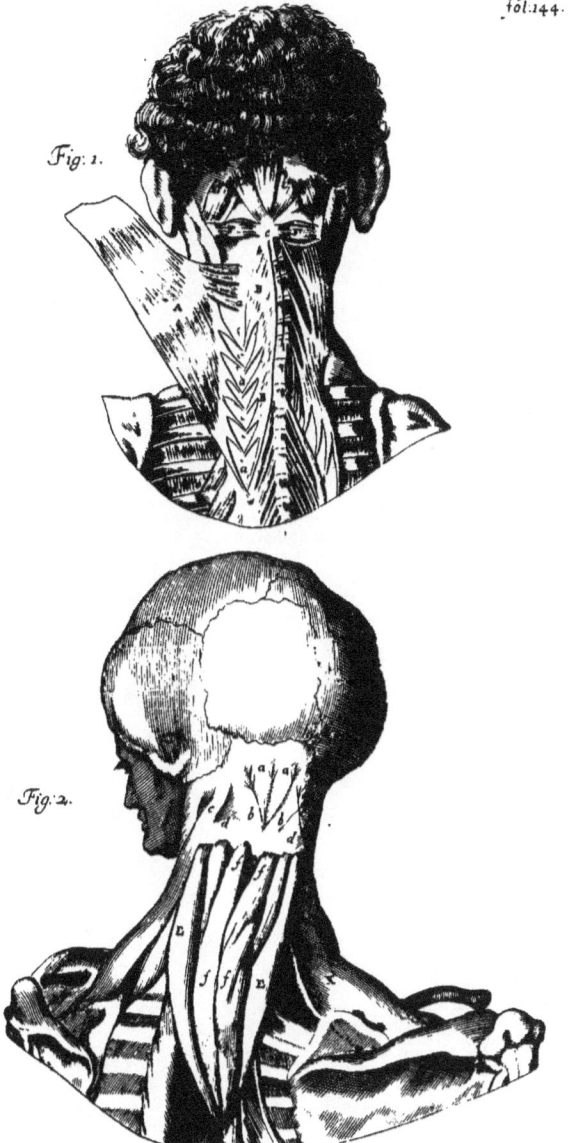

Fig. 1.

Fig. 2.

Longisſimus Dorſi.

THis Muſcle hath not this name given it, only becauſe it is planted between the Muſcles of the Back, but being as the longeſt of the whole Body : for it ariſeth from all the Spines of the *Os Sacrum*, and Vertebres of the Loins, as alſo from the inward part of the *Os Ileon*, where it joyns it ſelf to the *Sacrum*, being the firſt and chief pair, its beginning being externally ſtrong and Nervous, and ſomewhat Acute, but within Fleſhy, and ſo aſcending, doth fix it ſelf to the tranſverſe Proceſſes of the Loyns, and becomes more Fleſhy; then marching on, it narrows it ſelf, and gives a ſmall Nervous Tendon upon every tranſverſe Proceſs of the *Thorax*, except the twelfth, and doth inſert it ſelf into the ſaid Proceſs of the firſt Vertebre of the *Thorax*, although ſometimes it attains the Mammillary Proceſs.

This extends the Thorax.

This is allowed to extend both the *Thorax*, the Loyns, and their Vertebres : upon this borders the whole *Sacrolumbus* in its outſide, whoſe beginning with this is one and the ſame, and is alſo continued from the end of the *Os Sacrum* to the twelfth Vertebre of the *Thorax*, and ſo continued in its whole progreſs through the Loyns.

Uſe.

This you have at *H. H.* Fig. 1. Tab. 24. *I. I.* Shews the ſame laid bare in the Right ſide, *H. H. I.* You have the ſame at *Tab.* 26. *b. b.* The ſame laid bare.

Sacrolumbus.

This helps the former in extending the Thorax.

THis Muscle lieth under *Serratus Posticus Inferior*, having the same Origination with *Longissimus Dorsi*, and doth adhere to him externally lateral, according to its length, until it hath got the twelfth Vertebre of the *Thorax*, where they seem to be two, although scarce divisible by dissection, and so growing thinner, doth in'ert it self by two small Tendons into every Rib of the *Thorax* at their Incurvation.

Use.

About these Tendons there doth arise a great dispute amongst Anatomists; for some with *Laurentius* do think this Muscle to send forth a double Tendon, one upwards to the lower parts of the Ribs, the other downwards to their upper part; and these Tendons thus diversly mediating (which are manifestly seen about the Ribs) are seen to attol the Ribs upwards in inspiration, and to draw them downwards in expiration; The truth is, these contrary actions are not to be reasonably conjectured to be acted by one certain Muscle: and that it may very well be allowed, that these descending Tendons ought to proceed from some other peculiar Muscle, here upon a diligent enquiry, we find them to proceed from a certain Muscle substrated to the *Sacrolumbus*, and to this it is so closely annexed, that it is scarce perfectly to be separated from it.

This you have at *K. K. Tab.* 24. *L. M. N.* The same laid bare.

Cervicalis

Cervicalis Descendens.

This takes its Origination from the third, fourth, fifth, sixth, and seventh Vertebres of the Neck, and hence by *Diemerbroeck* is Christened by the name of *Cervicalis Descendens*, thence arising Fleshy, sending forth Tendons, somewhat downwards into the upper part of all the Ribs, directly opposite to those Tendons of the *Sacrolumbus*, and these Tendons thus intersecting themselves, do not work together but by turns; As the Tendons of the *Descendentis Cervicalis*, do draw the Ribs upwards in aspiration, so the Sacrolumbal Tendons in expiration do draw the Ribs downwards, so as they may be moved to, and contracted by them. *This doth extend the Thorax. Expiration.*

As to this Muscle the Author will further satisfie you, and therefore I recommend you to *Diemerbroeck*.

Sacer.

Sacer.

This extends the Loyns.

FOr the various Motions of the Back and Loyns, as forwards, backwards, and laterally, to every of the Vertebres are implanted Tendons of Muscles; Our Accurate Dissectors of late have found four pair of Muscles to be planted in the Back and Loyns, by whose help, the strong Motions of these parts are performed: some appointed for Contraction, others for Extension.

This pair are so called from their Origination, because they do arise from the Exterior part of the *Os Sacrum*, where they are joyned to the Spine, and so arising Fleshy, have three several Tendinous Insertions: the first into the upper part of the transverse Processes of the Vertebres of the Loyns, the other into the Root of the same Processes, and the third into the Spine of the same Vertebre; To find this out, you must raise *Latissimus Dorsi*, and *Sacrolumbus* from their Membranous Originations; at *Os Ileon*, *Os Sacrum*, and from the Lumbal Spines, and immediately under this will this appear *in situ*, these helping forwards the action of *Longissimus Dorsi*.

This you have at *L. M. N.* at *Tab.* 26. *L. M.* Shewing its beginning, *N.* Its Termination.

Semispina-

Semispinatus.

THis is the fourth pair, arising with a Nervous beginning *This extends the Thorax.* from all the Spines of the *Os Sacrum*, and Loyns, and becoming Fleshy, doth bestow a Nervous Tendon upon every Spine of the Vertebres ascending, and terminates acutely in the Spine of the first Vertebre of the *Thorax*, extending it.

This by *Spigelius* is accounted the second pair of the Muscles *Use.* extending the *Thorax*.

This is shewn at *Tab.* 26. at K. K. c. c. Shewing its beginning, d. Its Termination.

Quadratus.

This Muscle extends the Loyns.

THis Muscle is called *Quadratus* from the resemblance it hath with a square: it ariseth from the back part, and upper Cavity of *Os Ileon*, as also from the upper part of *Os Sacrum*, broad, thick and Fleshy, and so ascending over the Vertebres of the Loyns, doth adhere internally to their transverse Processes, and inserts it self being grown narrower, to the inner part of the twelfth Rib Fleshy, and doth bend the Vertebres of the Loyns forwards: and one only working, it bends it obliquely forwards to the sides.

Use.

Use. The Loyns are concerned with three Motions, the which are performed by two pair of Muscles, for they are bent forwards, extended backwards, and brought laterally, they are contracted by these two, and extended by *Sacer*.

This you have shewn at *R. R. Tab.* 26.

Psoas

Pſoas vel Lumbaris.

Since Ambulation is the proper office and work of the Foot, and this being very conſiſtent in Firmation and Motion, (for when one Foot remains on the Ground, the other is lifted up, and ſo moved forward for the compleating of Ambulation) now for the performance of either of theſe Functions, its very requiſite theſe parts ſhould be furniſhed with ſuch Muſcles as may make forwards towards theſe Extenſions and Contractions, as alſo to the promoting of the various Motions of the Thigh, Leg, and Foot, theſe being according to our pleaſures either extended, contracted, adduced, abduced, and wrought or brought about: amongſt the Contractors we meet with this, as the firſt called *Lumbaris* or ~~ʃ~~.

It ariſeth Livid and Fleſhy from the ſides of the Bodies of the two laſt Vertebres of the *Thorax*, and the three uppermoſt of the Loyns, and from their tranſverſe Proceſſes deſcending ſomewhat round from the inner part of the *Os Ileon*; to the *Os Pubis*, where it becomes a ſtrong and round Tendon, and running through its Sinues, is implanted into the upper part of the leſſer *Rotator*: Its Uſe is to draw the Thigh upwards, and to bend it inwards; and becauſe the Kidneys do lie upon this Muſcle, as *Laurentius* doth obſerve, over which is ſpread a notable Nerve, hence it happens, that ſuch as are troubled with the Stone, do find a ſleepineſs in that ſide of the Thigh whereon the Stone is lodged, by its compreſſion.

This bends the Thigh.

Uſe.

Obſ.

D D. Shews this at *Tab.* 27. *b. b.* Shews its Origination, *E.* Its Tendon, *F. F.* The Nerves which paſs hence into the Thigh, *G. G. G.* The ſame laid bare.

The Explanation of the Twenty sixth Table.

A. Splenius.
   a. a. a. Shews its Anſulæ.
B. B. Trigeminus.
C. *The ſame laid bare.*
D. D. Tranſverſales Colli.
E. Rectus Major.
F. Obliquus Superior.
G. Obliquus Inferior.
H. H. I. Dorſi Longiſſimus *laid bare.*
  b. b. Shews the ſame wholly laid bare.
  a. a. a. Shews its inward Face and Anſulæ.
K. K. Semiſpinatus.
L. M. N. Sacer.
L. L. *Shews its Origination.*
R. R. Lumbales Quadrati.

TAB. XXVI.

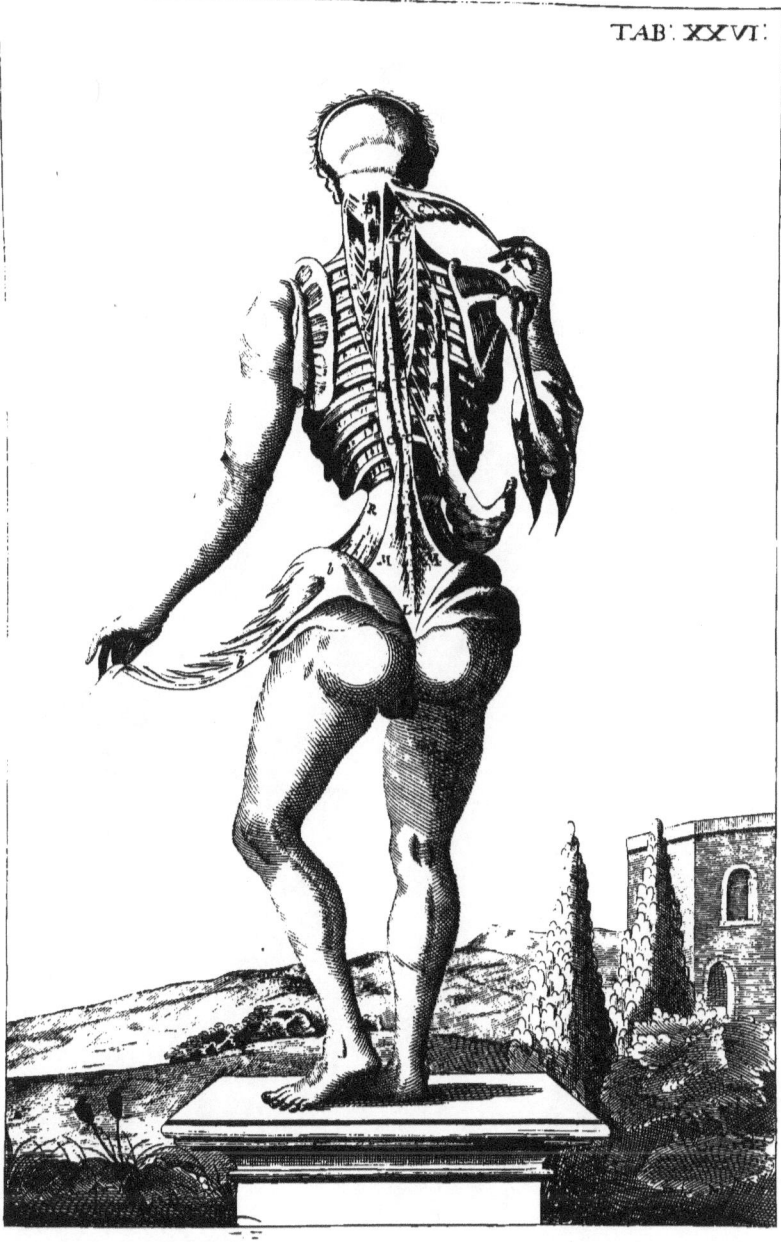

Pſoas Parvus.

Beſides the former, in ſome Bodies is ſeen this *Pſoas Parvus*, ſo called by *Bauhine*, it ariſeth Fleſhy the length of a little Finger, and is dilated with a ſlender and plain Tendon above the *Pſoas*, and ends with the *Pſoas* and *Ileon*, and embraceth them very firmly ; *Riolan.* affirms he never could find this in Women ; *Bartholine* writes, that he ſaw this Muſcle ariſing in a ſtrong and Fleſhy Man at the *Hague*, whoſe beginning was Fleſhy, and did equal the breadth of three tranſverſe Fingers ; it was inſerted Fleſhy in the upper Poſterior Margent of the *Os Ilii*, at the Origination of the *Iliacus Internus :* he ſuppoſed its Uſe to be as a Pillow to the former, and that whereas the *Os Ileon* of it ſelf was immovable, or that it might ſuſtain the *Os Ileon* erected, left by the ſtanding too much thereon, it might cauſe a wearineſs and prove burthenſom.

Our Maſter of Anatomy, Mr. *William Molins* in the Year of his being Maſter doth mention this Muſcle, in the Body then Diſſected by him.

This Muſcle lies under the former, but appears not very frequently.

This doth help the former.

Obſ.

Its Uſe.

The Explanation of the Twenty seventh Table.

A. *Shews* Quadrati.
   D. D. Pſoas *or* Lumbalis.
b. b. Shews its Origination.
E. Its Tendon.
G. G. G. The ſame laid ſomewhat bare.

I. I. I. Iliacus Internus.
L.L.M.M.M. The ſame laid bare.
O. Lividus.
P. The ſame laid bare.
R. Rectus.
T. Faſcialis.

We

TAB:XXVII

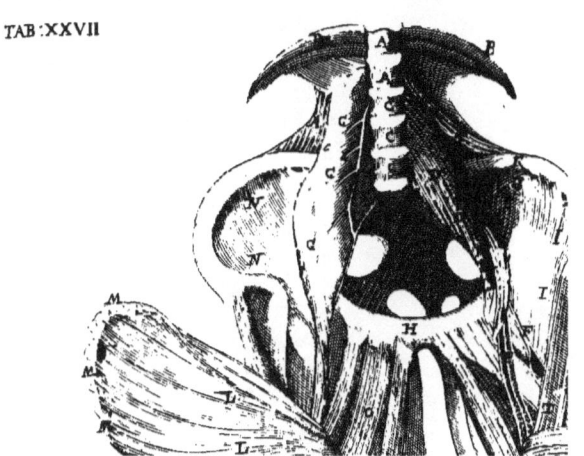

We arrive now to the taking the Thigh off from the Trunck of the Body, by dividing Os Ileon from the Os Sacrum, after which, will this appear as the most proper Order of Dissecting these following Muscles.

Iliacus Internus.

THis is the second Muscle of the Thigh, proceeding from the whole Spine of the *Os Ileon* semicircular broad, and internally Fleshy, then narrowing, and growing thicker becomes Tendinous, and passing through the same *Sinus* with the *Psoas*, is implanted into the same *Rotator* a little below it. This is allowed to bend the Thigh directly; that is, when it is raised towards the Body, so as that it inclines to no one side, and when we make a right Angle with the Spine, then we properly say the Thigh is contracted, when sitting we sit with contracted Thighs.

This contracts the Thigh directly.

Use.

This you have at 27. *Tab.* at *I. I. I. L. L.* The same laid bare *id. Tab. M. M. M.* Shews its Fleshy beginning.

Glutæus

Glutæus Major.

This extends the Thigh obliquely backwards.

THis is the first of the Extenders, the which with its other two doth make up the Fleshy Mass of the Buttocks, the Skin being laid bare, this shews its broad beginning, enated from diverse Bones: it ariseth from the whole Spine of the *Os Ileon* externally, then from the lower part of the *Os Sacrum* laterally; And thirdly, from the *Os Coxendix* large, and Fleshy, running obliquely down over the Juncture of the *Os Coxendix*; and growing narrower is implanted by a broad and strong Tendon into the first Impression of the great *Rotator*, and part of it also into the *Linea aspera*.

Use.
Then we properly do affirm the Thigh to be extended, when it's brought outwards, (*viz.*) as when we stand as it were with divaricate Thighs, or Thighs that are placed at a distance.

This you have at *Tab.* 28. at *C. C. C. a. a. a.* Shews its upper part, *b.b.* Its other part, *H. I. K.* The same laid bare. *H. H. H.* The thick and Fleshy beginning of it, 1.1.1. Its thick Belly, *K. K.* Its Tendinous Substance.

Psoas

Glutæus Medius.

THis Muscle ariseth under the former, much like it both *This extends it obliquely* in *site* and magnitude, from the forepart of the Spine, *forwards.* as also from the back of *Os Ileon*, Fleshy, broad, and semicircular, and obliquely descending, narrows it self, and doth enwrap the Juncture as the former, and is implanted by a broad, strong and Membranous Tendon transversly into the fourth impression of the great *Rotator*; This is said to extend *Use.* the Thigh and draw it upwards, and laterally forwards, as some imagine, but I humbly conceive it assisteth the *Glutæus Major*, and depresseth the *Os Femoris*: and after it is lifted up by the *Psoas* and *Iliacus Internus*, this Muscle pulleth it down again, and is as an Auxiliary Muscle to the *Glutæus Major*, and *Minor*, in the extension of the Thigh.

L. L. Shews this Muscle *in situ* at *Tab.* 28. *e. e.* Shews its Fleshy beginning, B. B. Shews the same *in situ*, *Tab.* 29. *D. e. f.* Shews it laid bare, *D. D. D.* Shews its Fleshy beginning, *E*. Its Fleshy Belly, *F*. Its Tendon.

Glutæus Minor.

This extends the Thigh directly.

THis lies wholly under the second, arising livid, broad, Semicircular and Fleshy, about the lower part of the back of the *Ileon* near the *Acetabulum* of the *Coxendix*, and so runs obliquely forwards Fibrous, according to the Ligament that binds in the head of the *Os Femoris*, and is implanted by a broad and strong Tendon into the third impression of the great *Rotator*.

Use.

These three do extend the Thigh, and do draw it backwards, and so extend it. I conceive it most probable, they all unite in pulling the *Os Femeris* downward and backward, after it is elevated by the *Flexors*, the *Psoas*, and *Iliacus Internus*.

This is shewn at *Tab.* 29. at G. G. *a. a.* Shews its beginning, This you have laid bare at *Tab.* 29. at *a. a. a. b. b.*

Iliacus

Iliacus Externus vel Pyriformis.

THis Circumagent Motion is performed when the Thigh is Circumverted; which thus happens: when standing with the Right Foot firm upon the Ground, we move the Thigh obliquely; this Motion is twofold, for its either brought outwards or inwards: outwards when the Knee is moved outwards, and the Heel inwards: inwards when the Knee is brought to the Ham, and the Heel forwards; Amongst these Circumagent Muscles this is reckoned the first *This brings the Thigh about upwards.*

It ariseth thick and Fleshy from the three lower Vertebres of the *Os Sacrum*, and running transversly becomes a round Tendon, and is inserted into the fourth impression of the great *Rotator*; This draws the Thigh upwards, outwards, and backwards, this is allowed the thickest Muscle in Human Body, and is stuffed out with wonderful variety of Fibres. *use.*

This is shewn at *Tab.* 28. at *M. f.* Shews its beginning, *g.* Its Tendon, *H. Tab.* 29. Shews the same *in situ*, *K.* Shews the same laid bare, *L.* Shews its beginning, *M.* Its long and slender Tendon, This you have laid bare at *Tab.* 32. at *L.*

The

The Explanation of the Twenty eighth Table.

C. C. Glutæus Magnus.
a a. a. Shews the Circumference of its Origination.
H. I. K. The same laid bare.
H. H. H. Shews its first and Fleshy Origination.

I. I. I. Its Venter.
K. K. Its Tendinous Substance.
L. L. Glutæus Medius.
C. C. Shews its Fleshy beginning.
M. Pyriformis.
Q. Q. Marsupialis.

Obtura-

TAB. XXVIII.

Obturator Internus five Marfupialis.

THis ariseth Fleshy and large from the Membrane inter- *This brings the Thigh round about outwards.*
nally that covers the great perforation of the *Os Pubis*,
and covering the whole inward face of that Bone and
Coxendix, grows narrower, and fends forth three or four Tendons
which are carried through the *Sinus* of the *Coxendix*, which is
arched over, according to its length, with a strong Ligament,
backwards to the outward part of the *Coxendix*, where they are
received into a Fleshy Purse, and so making one Tendon, are
implanted into the *Sinus* of the great *Rotator*, and doth make
the outward Rotation; This Muscle must be raised inwards,
and got through the *Sinus* under the Ligament, then its Purse
will plainly appear.

This is shewn at *Tab.* 28. at Q. Q. O. Shews the same at *Tab.* 29.
This you have also, *Tab.* 29. at D. E. F. G.G. Shews the same
laid bare, *H.* Shews its Fleshy Purse.

Quadrigeminus.

This brings the Thigh about backwards.

THis ariseth Fleshy from the rising of the *Os Ileon*, and from the Appendix of the *Coxendix*, and runs broad, short and Fleshy towards the hinder part of the great *Rotator*, and is inserted into that space of the Bone which is between the two *Rotators*; the head of *Iividus* and part of *Triceps* must be thrown off, before the Origination of this will be cleared, or *Obturator Externus* found out.

This Muscle by *Vesalius* is divided into two Muscles.

Obs.

These Circumagent Muscles do then bring the Thigh about, when standing directly, and firm on the Earth we move the Thigh obliquely, and this Motion appears double, as outwards or inwards; that properly granted inwards, when the Knee is brought toward the *Poples*, and the Heel carried outwards: that outwards, when the Knee is carried outwards, and the Heel brought inwards.

This you have at *Tab.* 30. at *I*.

Obtura

Obturator Externus.

This hath its name from its Origination, it arising from the outward part of the Cavity, and is subjacent to the *Pectinæus*, it arising large and Fleshy from the Membrane that enwrappeth the perforation of the *Os Pubis* externally, and so running transversly to the back part of the Thigh, becomes narrower, and is inserted by a strong Tendon into the the *Sinus* of the great *Rotator*, and doth direct the inward Rotation. You must carefully bring your Knife inwardly about the edge of the perforation of the *Os Pubis*, and it will both arise and appear the better.

This brings the Thigh about inwards.

Use.

This (together with the *Internus*) fill up the Cavity which lies betwixt the *Os Pubis* and *Ischium*, whence it has its denomination.

This you have at *Tab.* 32. at *H. I. K.* Shews the same laid bare, *C. C.* Shews the same at *Tab.* 30. *S.* Shews the same at *Tab.* 31. This you have also at *Tab.* 32. at *H.* in its place, *I. K.* Shews it laid bare.

The Explanation of the Twenty ninth Table.

A. *A. A.* Lividus.
C. C. Obturator Externus.
D. D. E. Obturator Internus.
D. D. *Shews its broad and Semicircular head*
E. E. *The Tendons thereof.*
F. F. *The Marſupium or Purſe it ſelf*
G. G. *Shews the ſame laid bare*
H. *Shews its Fleſhy Purſe.*
Q. Quadrigeminus.
K. Seminervoſus.
L. Semimembranoſus.

M. Biceps.
O. Seminervoſus *laid bare.*
P. *Shews its Fleſhy Origination.*
Q. *Its Nervous Termination.*
R. Semimembranoſus *laid bare.*
S. *Shews its Nervous head.*
V. Biceps *laid bare.*
X. *Shews its Origination.*
Y. *Shews its Tendinous Inſertion.*
a. a. a. b. b. Glutæus Minimus *laid bare.*
c. Pyriformis *laid bare.*
m. m. m. *Shews* Triceps *in* ſitu.

Membra

Tab. XXVIIII

Membranosus.

THe Leg hath three motions allowed it, it being either *This exten-* extended, contracted, or brought somewhat obliquely *the Leg di-rectly.* outwards, all the Extenders are implanted in the fore side of the Thigh, and these working together, do extend it aright; This Muscle by Anatomists is reckoned as the first of the Extenders.

It ariseth sharp, externally Nervous, inwardly Fleshy, from the Spine of the *Os Ileon*, on that side next *Sartorius*, and then becomes broad, and Nervously Membranous, enwrapping all the Muscles of the Thigh within its self; then covering the Patella and two Focills in their outward part, is there implanted, and doth extend the Leg directly; and as some Authors will *Use.* have, doth somewhat abduce it outwards.

As much as possibly may be, is to be kept of this Membrane in Dissection, and the division thereof to be made in the back part.

This yon have at *Tab.* 30. at *M. M.* 0.0.0. *M. M.* Shewing its beginning, 0.0. Declaring its broad Tendon, Q R. S. Shews the same laid bare.

Sartorius five Fascialis.

This bends the Leg.

THis Muscle hath its name from its daily use which is made of it by Taylors, and Shoomakers, who when they be at their work, do generally sit Cross-leg'd, some call this *Fascialis* from *Fascia*, a Ligature or Swadling Band.

It ariseth sharp, Fleshy and Nervous from the fore part of the Spine of the *Os Ileon*, and then becoming Fleshy and broad, runneth obliquely internally over the Muscles of the Thigh, becoming Tendinous and broad at the inward Appendix of the *Os Femoris*; and is implanted by a broad Tendon, as some Authors affirm, and round, as others into the *Tibia*.

Use.

Riolan. writes, that this doth not bend the Leg, but rather doth bring it inwards, and so he supposeth doth more aptly extend it; This is one of the longest Muscles in Human Body.

This you have at *Tab.* 30. at *A. A. B.* Shews its Origination, *C.* Its Termination, *D. D.* Shews the same laid bare, *T.* Shews it also at *Tab* 27.

Gracilis.

Gracilis.

THis is generally accounted the second of the Contractors: *This doth assist the former.* it ariseth large and Nervous from the middle of the *Os Pubis*, according to the length of its Cartilage, and so descending inwards towards the Ham Fleshy, doth there become a round Tendon, at the inner head of the *Os Femoris*, inserting it self into the *Tibia* near the former.

The first and second of the Flexors are planted in the fore *obs.* part of the Thigh, the one outwards, the other inwards, the rest in the back part thereof; if they be all contracted together, they do bend the Leg directly, but the first four only working, they do bring it somewhat inwards, whilst the other endeavours to carry it outwards, and by this motion, the end of the Foot in which the Toes are implanted is somewhat brought outwards.

This is shewn at *F. F.* in its place, *Tab.* 30. *G.* Shews its Commissure, *H.* Shews its Insertion, *I. K. L.* Shews the same laid bare, *I.* Shewing its Nervous beginning, *K.* Its round Belly, and *L.* Its round Tendon.

The Explanation of the Thirtieth Table.

A. *Fascialis or* Sartorius.
D. D. *Shew the same laid bare.*
F. F. Gracilis.
I. K. L. *The same laid bare.*
M. M. O. O. Membranosus.
Q. R. S. *The same Muscle laid bare.*
V. V. Triceps *in either side.*
X. Obturator Externus.
Y. Lividus.
Z. *The same laid bare.*
a. a. Rectus in situ.
b. b. Vastus Internus in situ.

Rectus.

Fol. 162

TAB. XXX.

Rectus.

THis hath its name from its right Progress, and hath the *This extends the Leg.* Figure of a true Muscle, and is held as the third of the Extenders: it ariseth sharp and Nervous from the small inner Extuberance of the *Os Ileon*, and then becoming Fleshy and round, when it arrives at the *Patella* it expands it self into a strong broad Tendon, entirely covering it, and running downwards, is inserted into the outward part of the *Tibia*, at a prominency provided for it.

This you have at *A. A. Tab.* 31. *B.* Shews the inward protuberance of the *Os Ileon* whence it ariseth, *C.* Shews its Tendon, *D. E. F.* The same laid bare, *D.* Shews its sharp and Nervous beginning, *E.* Its Fleshy Venter, *F.* Its strong Tendon; This you have also at *Tab.* 30. at *a. a. a. a. K.* Shews this also at *Tab.* 27.

X x   Vastus

Vaſtus Externus.

This also doth extend the Leg.

THis from its great Maſs of Fleſh, hath its name given it, and is the third of the Extenders, ariſing broad and Nervous from the Root of the great *Rotator*, cleaving to the upper and outward part of the *Os Femoris*, and ſo deſcending Fleſhy to the *Patella*, it becomes a Membranous broad Tendon, and mixing with the Tendon of the *Rectus* makes the ſame covering for the *Patella*, carrying the ſame inſertion with it.

This you have at *G. G. G. G.* in its place at *Tab.* 31. *H.* Shews the beginning of this Muſcle, *I.* Its Tendinous Membrane.

Vaſtus

Vaſtus Internus.

THis is the fourth extending Muſcle, ariſing Nervous from part of the leſſer *Rotator*, and from the neck of the *Os Femoris*, and growing Fleſhy, adheres to the Anterior, and Interior part of the ſaid *Os Femoris*, and recovering the *Patella*, mixeth it ſelf with the two former, covering the *Patella*, and taketh its inſertion in the ſame place as they do. *This works as the former.*

This you have at *L. L. L. Tab.* 31. This you have alſo in its place at *Tab.* 30. at *v. v.*

Biceps.

Biceps.

This doth contract the Leg.

THis is the fifth inflecting Muscle, arising sharp and Nervous from the Appendix of the *Coxendix*, and growing Fleshy, runs down externally, and being got half way, it attenuates and grows Nervous, as tending to a Tendon, where it joyns it self with its other head which ariseth from the *Os Femoris*, where *Glutæus Major* hath its insertion, and so growing thicker, outwardly Nervous, becomes a strong Tendon, and runs through the outward *Sinus* of the outward part of the head of the *Os Femoris*, and fixeth it self firmly to the outward side of the upper Appendix of the *Fibula*;

o.sf. Sometimes this Muscle is distinguished with a double beginning and ending, so observed by *Vesalius*, and therefore by him this is called *Duplex*.

Uf. This Muscle doth bend the Leg and pull it backward.

This you have laid bare at *Tab.* 32. at *O. O.* At *Tab.* 29. you have it at *M. V.* Shews this laid bare at *Tab.* 29. X. Shews its beginning, *Y.* Shews its Tendinous Substance, Z. Its Fleshy Mole or Substance.

Semi.

Semimembranosus.

This is accounted the fourth of the Inflectors, arising *This bends the Leg.* where the former did, with a small Membranous beginning, and running downwards by the back part of the Thigh, doth continue Membranous half its progress; and then becoming Fleshy and thick, inserteth it self by a round Tendon into the Ham, in the inner side of the *Tibia*; The first of the four Inflectors being tied into the inner side of the *Tibia*, and the fifth into the *Fibula*; the first makes the inward hardness in the Ham, the other, the outward.

This you have at *Tab.* 29. at *L*. You have the same laid bare at *R*. *S*. Shews its Nervous Head, *T* Its broad Tendon.

The Explanation of the One and thirtieth Table.

A. Rectus.
D. E. F. The same laid bare.
D. Shewing its sharp and Nervous Orination.
E. Its Fleshy Venter.
F. Its strong Tendon.

G. G. G. G. Vastus Externus.
H. Its beginning.
I. Its Membranous Tendon.
L. L. L. Vastus Internus.
Q. Q. Q. Q. Triceps *in either side.*
S. S. Obturator Externus.

Seminer-

Fol. 166

TAB XXXI

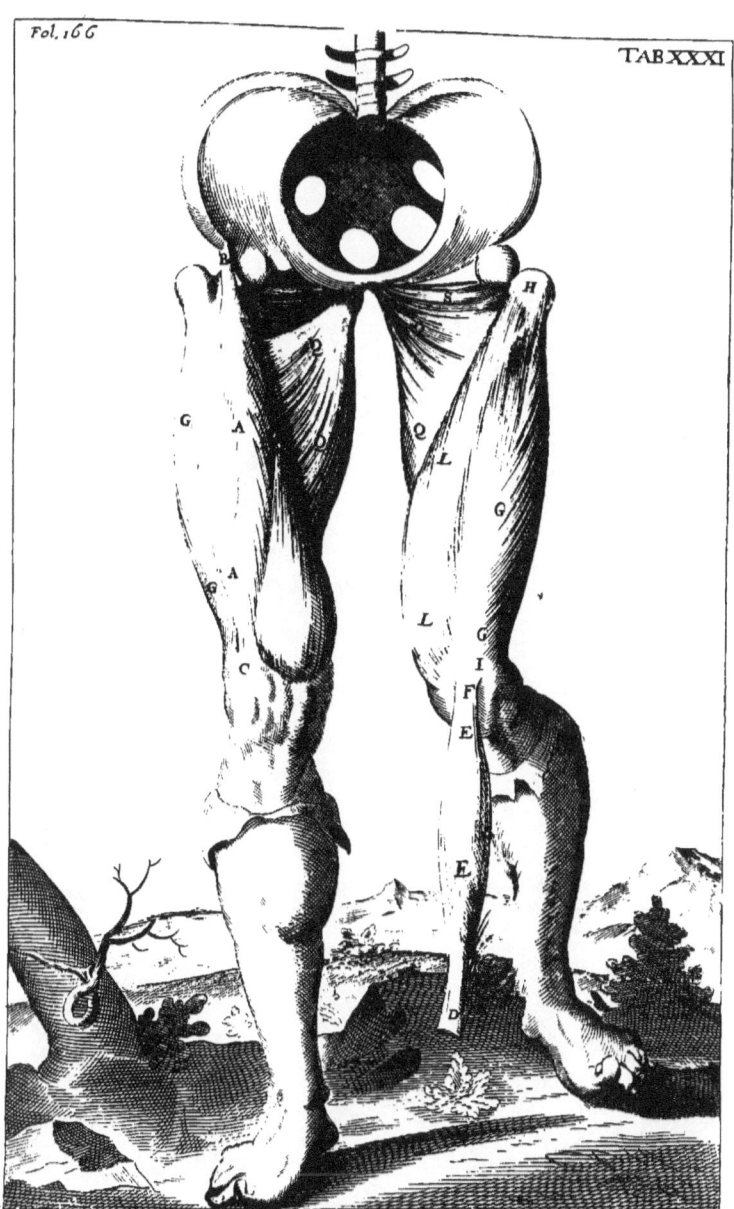

Seminervosus.

THis hath its name from its Substance, it being partly *This works*
Nervous, and partly Fleshy, and is the third pair: it *as the for-mer.*
ariseth small and Nervous from the same Appendix as
the former, and so continuing half way in its descent, it then
becomes Fleshy, running by the back part of *Os Femoris*, to
the Ham, near which it becomes a round Tendon, and reflect-
ing it self, is inserted into the forepart of the *Tibia*.

This Tendon hath allowed it this worth observation, that it *Obs.*
reacheth even to the middle of the length of the *Tibia*, with the
rest of the Tendons implanted to the *Tibia*, the which do scarse
descend so far.

This you have at *Tab.* 29. at *K*. *O*. Shews the same laid bare,
P. *P*. Shews its Nervous beginning, *Q*. Its Nervous Tendon.

Triceps.

Triceps.

This brings the Thigh inwards.

THis is the largest of all the Muscles of the Thigh, yea, I may say, of the whole Body; its apparently seen to have three heads, all which do conclude in one end: It ariseth with three heads, the first Fleshy and Nervous from the Appendix of the *Coxendix*, the which swelling, doth dilate it self into the hinder part of the Thigh, and then growing small, doth end in a round Tendon, at the inner head of the said *Os Femoris*; The second ariseth Fleshy from the *Coxendix* at its conjunction with the *Pubis*, and terminates at the Root of the lesser *Rotator*, and in the upper part of the *Aspera Linea*; The third ariseth Fleshy from the lower part of the *Os Coxendix*, and is implanted into the *Linea Aspera* of the said *Os Femoris*; To which some add a fourth, (*viz.* the following call'd *Pectineus*) which seems to be a part of this.

Use.

This is a Riding Muscle, drawing the Thigh inward, and fixeth the Rider to his Seat, keeping him firm in the Saddle, and may be truely stiled *Musculus Pudicitiæ*, as assisted by the *Lividus* in keeping the Legs close.

This you have at *Tab.* 31. at *Q. Q. Q. Q.* *A. A A. A.* Shews it in *Tab.* 32. *B.* Shews its beginning, *D. D. D.* Shews the same laid bare, *F.* Shews its beginning, *F. G.* The division of its Tendon.

Lividus

Lividus five Pectinalis.

THis is allowed the fourth of the Inflectors, it ariseth broad and Fleshy from the forepart of the *Os Pubis*, near its Cartilage, obliquely descending, and is inserted by a large and short Tendon to the inside of the *Os Femoris*, near its middle, so as it strongly brings the Thigh upwards and inwards. *This bends the Thigh obliquely inward.*

By *Bartholine* this Muscle is reckoned amongst the Adducent Muscles.

This Muscle is assistent to the *Triceps*, being an Adductor of the Thigh, which it pulleth inward, being very useful in Riding, keepeth a Horseman close to the Saddle; and (as was before said) seems to be a part of the *Triceps*, though it does not so closely adhere to it, but it may be separated without difficulty. *Use.*

This you have at *Tab.* 29. at *A. A. A. A. B.* Shews the forepart of the *Os Pubis*, from whence this Muscle takes its Origination, *O.* Shews the same at *Tab.* 27.

The Explanation of the Two and thirtieth Table.

A *A A A.* Triceps
B Shews its Origination.
D. D. D. The same Muscle laid bare.
H. Obturator Externus.

I. The same laid bare.
K. Shews its Tendon.
L Pyriformis *laid bare.*
O O. Biceps *laid bare.*

Fol. 171 TAB XXXII

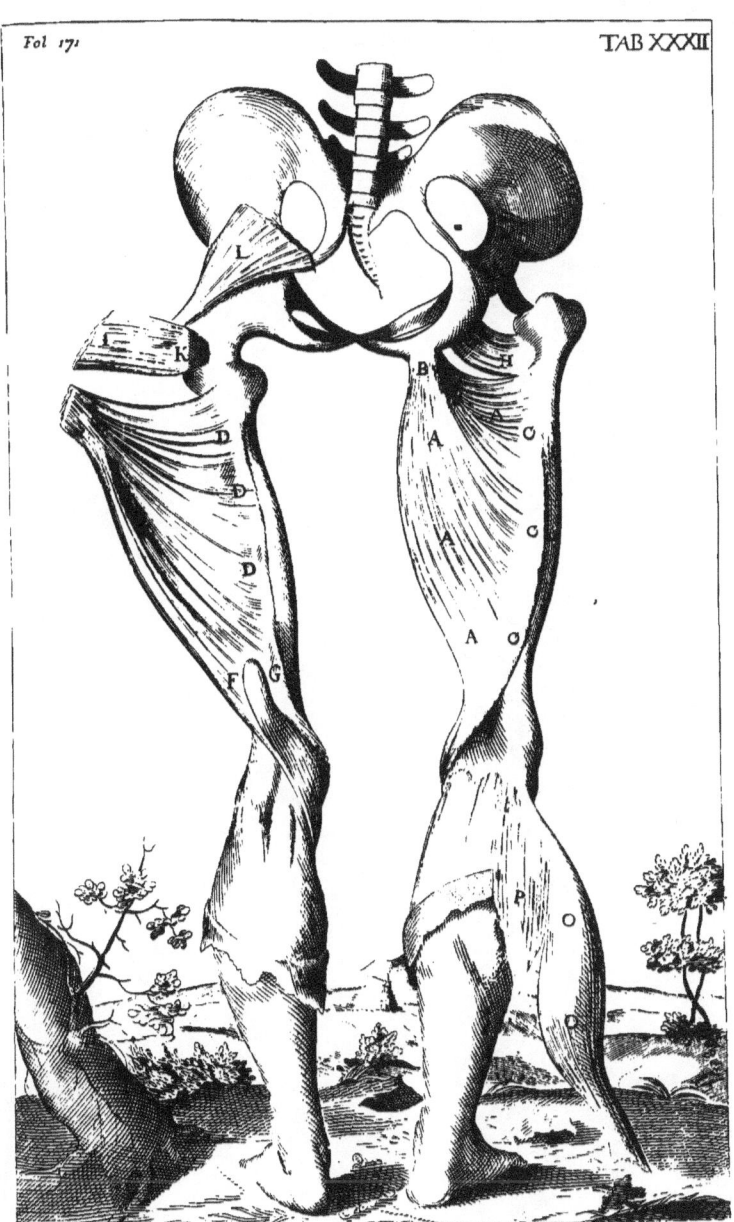

Gasterocnemius sive Gemellus Externus.

THe Foot hath allowed its threefold motion, (*viz.*) Ex- *This extends the Foot.* tension, Contraction, and lateral Motion; this is accounted the first of the Extenders, which maketh the Calf of the Leg: it ariseth broad and Fleshy from the inner head of *Os Femoris*, as also from the outward head of the same Bones; so descending according to their Originations, they are united about midway, and are converted into one entire, broad, strong, and Nervous Tendon, and becometh one with the Tendon of *Gasterocnemius Internus*, and doth insert it self into the back part of *Os Calcis*, so that indeed they are but one Muscle with a double Origination.

Vesalius doth assert that the Sesamoidal Bones are Tributary to the two heads of this Muscle, not far from their Origination.

This you have at *Tab.* 33. at *D. E. F. G. H. Fig.* 1. *D.* Shews its first Origination, *E.* Another of its beginnings, *F.* Its Coherence, *G. G.* Shews its large Fleshy Belly, *H.* Shews its large Tendon, at *Fig.* 2. *ejusd. Tab. O. P. Q. S.* Shews the same laid bare.

Planta-

Plantaris.

This moves the Skin of the Sole of the Foot.

AS that Muscle Is called *Palmaris*, which with its long and round Tendon doth march through the whole Interior part of the *Cubite*, and from thence to the Annular Ligament of the *Carpus*, and is afterwards expanded into a broad Tendon covering the whole Palm of the Hand; so also is this *Plantaris* expanded through the whole hinder part of the *Tibia*, with a long and round Tendon, and at length marcheth into the Sole of the Foot.

It ariseth Fleshy, round, and slender under the former, from the outward lower head of *Os Femoris*, and after some Dilatation, it becomes a slender round Tendon, and running obliquely from the outward between both the *Gasterocnemii* inwards, and being joyned with their Tendons near the Heel laterally, running forwards, doth insert it self (after Dilatation over the Sole of the Foot) into each of the five. Toes at the first Joynt.

Note. The Dilatation of this Muscle over the Sole of the Foot, as also its Insertions, will not be found, unless you allow it to insert it self into *Flexor Primi Internodii* at its Origination in the Cavity of the *Os Calcis*.

Use. This Muscle is of as great service to the Foot as *Palmaris* is to the Hand; and serves to the Extention, or Expansive Motion of the Foot.

This you have at *Tab.* 3. *Fig.* 2. at *I. K. K. I.* Shewing its head, *K. K. K.* Its round Tendon.

Gasterocnemius Internus, seu Soleus.

This ariseth livid, strong, and Nervous from the Posteriour Appendix of the *Fibula*, and growing larger, adheres both to that and *Tibia*, and descending at half way, it becomes narrower and Tendinous, making one with the *Gasterocnemius Externus* both in Origination and Insertion; These three Muscles are united about their ends, and do frame one very strong Tendon, implanted in the back part of the Heel, the which by reason of its greatness, and singular strength above the Tendons of other Muscles, obtains the name of *Chorda Magna*, the which being Bruised or Wounded, (as *Hipp.* writes) proves Mortal or very dangerous: its by this Tendon at this very day that our Butchers do daily hang up their Oxen by; And that which is worthy observation as touching this, is, that if any Inflammation doth arise about this part, it brings the whole Body miserably into consent therewith. *This extends the Ancle. Obs.*

This you have at L. M. N. Fig. 2. Tab. 33. L. Shewing its strong and Nervous head, M. Its large venter, N. Declaring its Tendon.

Suppopliteus.

This moves the Leg obliquely.

THis ariseth broad and Nervous from the outward head of *Os Femoris*, and growing Fleshy, runs obliquely to the back and inward part of the upper Appendix of the *Tibia*, and is there implanted.

Use.

Riolan faith, he sometimes hath found this double; This moves the Leg obliquely outwards, and turns the Foot somewhat inward towards the other.

This you have at *G. H. Tab.* 33. *Fig.* 2. *G.* Shewing its beginning, *H.* Its Termination.

Flexor

(187)

Flexor Digitorum Tertii Internodii, seu Perforans, sive Sublimis.

THis ariseth Fleshy, and long from the back part of the *Tibia*, running, and adhering according to its length, to the middle of it, and there becoming a Tendon, is carried to the Internal *Matleolus*, where it becomes round, and is carried under the Ligament that proceeds from the lower Appendix of the *Tibia*, to the *Os Calcis*, and then divides it self into four Tendons, which terminates at the third Joynt of the four lesser Toes. *This bends the Toes in the third Joynts.*

Obj. The Toes of the Feet are both contracted, extended, and moved laterally; And for these three Motions, Nature hath designed three kind of Muscles, and these are called either Flexors, Extensors, or Oblique Movers; four of these are implanted in the Leg; as this first, the Flexor of the Great Toe, and the Extensor of the third Joynt of the Toe, and the *Extensor Pollicis*; the other in the Foot.

L. L. L. L. Shews the four Tendons of this Muscle at *Tab.* 37. *Fig.* 2. *B.* Shews the Tendon of this Muscle, *Tab.* 37. *Fig.* 2. *D. D. D.* Shews its Fleshy part, *E. E. E. E.* Shews its four Tendons, *F. F. F. F.* Shews these also and their Originations at *Tab.* 37. *Fig.* 2. *H.* Shews this also at *Tab.* 34. *Fig.* 1. *I.* Shews its middle part, *K.* Its Exquisite Tendon, *D. D. E.* Shews the same, *Fig.* 2. *id. Tab.*

The

The Explanation of the Three and thirtieth Table.

FIG. I.

D E. F. G. H. I. Gasterocnemius Externus.
D. *Shews i s first Origination.*
E. *A second of its Originations.*
F. *Its Coition or Connexion.*
G.G. *Its large Fleshy Belly.*
K. *Its Tendinous Insertion.*

FIG. II.

G. H. Suppopliteus.

G. *Shews its Origination.*
H. *Its Termination.*
I. K. K Plantaris.
I. *Shews its head.*
K. K. K. *Its Tendon.*
M. N. N. Gasterocnemius Internus.
O. P. Q. *Shews* Gasterocnemius Externus *laid bare.*

Tibiæus

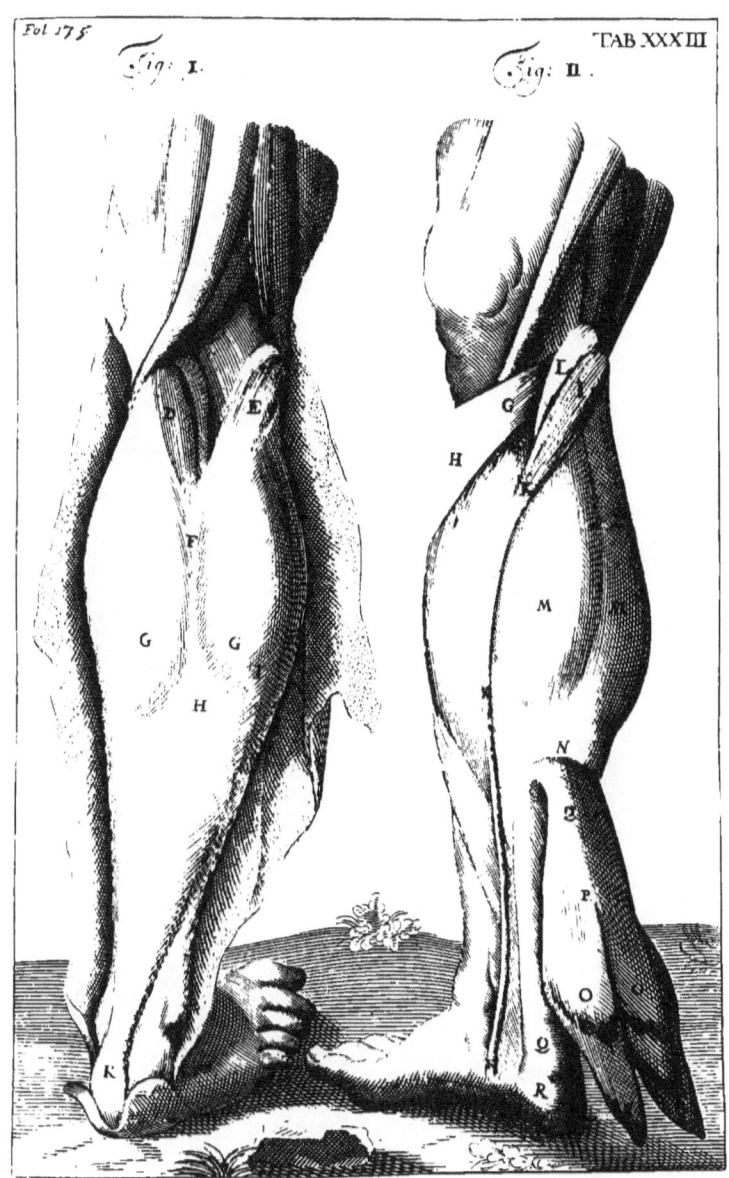

Tibiæus Posticus seu Nauticus.

IT ariseth Fleshy from the Ligament which joyns the *Tibia* and *Fibula*, according to their length, and from both Bones backwards, becomes a round Tendon, near the *Malleolus Internus*, where being bound by a strong Ligament, it overspreads it, and recovering the Sole of the Foot, is inserted into that lower part of that *Os Tarsi*, which joyns it self with *Os Cubiforme*; Sometimes there is seen, that it hath produced two Tendons, the one implanted into the *Os Naviculare*, the other into the Innominated Bone. *This brings the Foot inwards.*

This Muscle is called *Nauticus*, from the use which Seamen make of it, when they do run up the Shrouds. *vs.*

This you have at *Tab.* 35. *Fig.* 1. at E. E. *e.* Shewing its Fleshy beginning, *f.* Declaring its Tendon, D. D. D. Shews the same at *Tab.* 36. *Fig.* 2. *E.* Shewing its Tendon.

Flexor Pollicis.

This bends the Great Toe.

THe Great Toe hath allowed it variety of Muscles, the first of which is this arising sharp and Fleshy about the middle of the back part of the *Fibula*, descending larger, nearer the inward *Malleolus*, running obliquely under it, and is inserted into the last Bone of the Great Toe; Under this Ligament lodgeth the *Os Sesamoides*.

Obs.

This Tendon is seen sometimes to conjoyn with the Tendon bending the third Joynt, running to the second Toe, and before it reacheth the second Joynt of the Great Toe, requires the largest *Os Sesamoides*, which is in the other Joynts of the Toes.

This you have at *Tab.* 37. *Fig.* 2. *c* Shewing its Tendon, *D. D. D.* Shews its Fleshy part, *Tab.* 36. *Fig.* 1. at *N.* you have it laid bare, *M.* Shews this at *Tab.* 34. *Fig.* 1. *N.* Shews its long and narrow Tendon, *I. I.* Shews the same exactly at *Tab ejusd. Fig.* 2. *K. K.* Shews its Tendon, *N.* Shews this laid bare at *Tab.* 36. *Fig.* 2.

Flexor Secundi Internodii, Perforatus seu Profundus.

THis is a second of the Inflectors of the Toes, by some called *Flexor Brevis*, it ariseth Fleshy and Membranous from the extremity of *Os Calcis*, and marching half way the Sole of the Foot, divides it self into four round Tendons, which at their Insertions into the second Joynts of the lesser Toes are perforated for the transmission of the *Tertii Internodii Flexor*. *This bends the Toes in the second Joynt.*

This you have at *Tab*. 37. *Fig*. 1. at *C. C. C. D. D. D.* Shewing its four Tendons; And at *Fig*. 2. *ejusd. Tab*. you have the same laid bare at *N. O*. Shewing its Origination, *P. P. P. P.* Declaring its Tendons, *Q*. Shews the same laid bare at *Tab*. 34. *Fig*. 2: *N*. Shews the same laid bare at *Tab*. 37. *Fig*. 2. *O*. Shews its Origination, *P. P. P. P.* Shews the four Tendons of this Muscle.

The

The Explanation of the Thirty fourth Table.

FIG. I.

F G. Suppopliteus *laid bare*.
F. *Shews its Fleſhy beginning*.
G. *Its Fleſhy Termination*.
H. Flexor Tertii Internodii Digitorum.
I. *Shews its length*.
K. *Its exquiſite Tendon*.
L. L. Peroneus Primus.
M. Flexor Pollicis.
O. Pollicem Adducens in ſitu.
P. P. Gaſterocnemius Internus *laid bare*.
Q. Plantaris *laid bare*.
R. R. R. *Shews its long and round Tendon*.

FIG. II.

D. E. E. Flexor Tertii Internodii Digitorum.
D. *Shews its ſharp Origination*.
E. E. E. *Its ſlender and Fleſhy Venter*.
F. *Its Tendon*.
I. I. Pollicis Flexor.
K. K. *Shews its Tendon*.
L. L. Peroneus Primus in ſitu.
M. *Its Tendon*.
P. Minimum Digitum Abducens.
Q. Pollicem Adducens *laid bare*.

Flexores

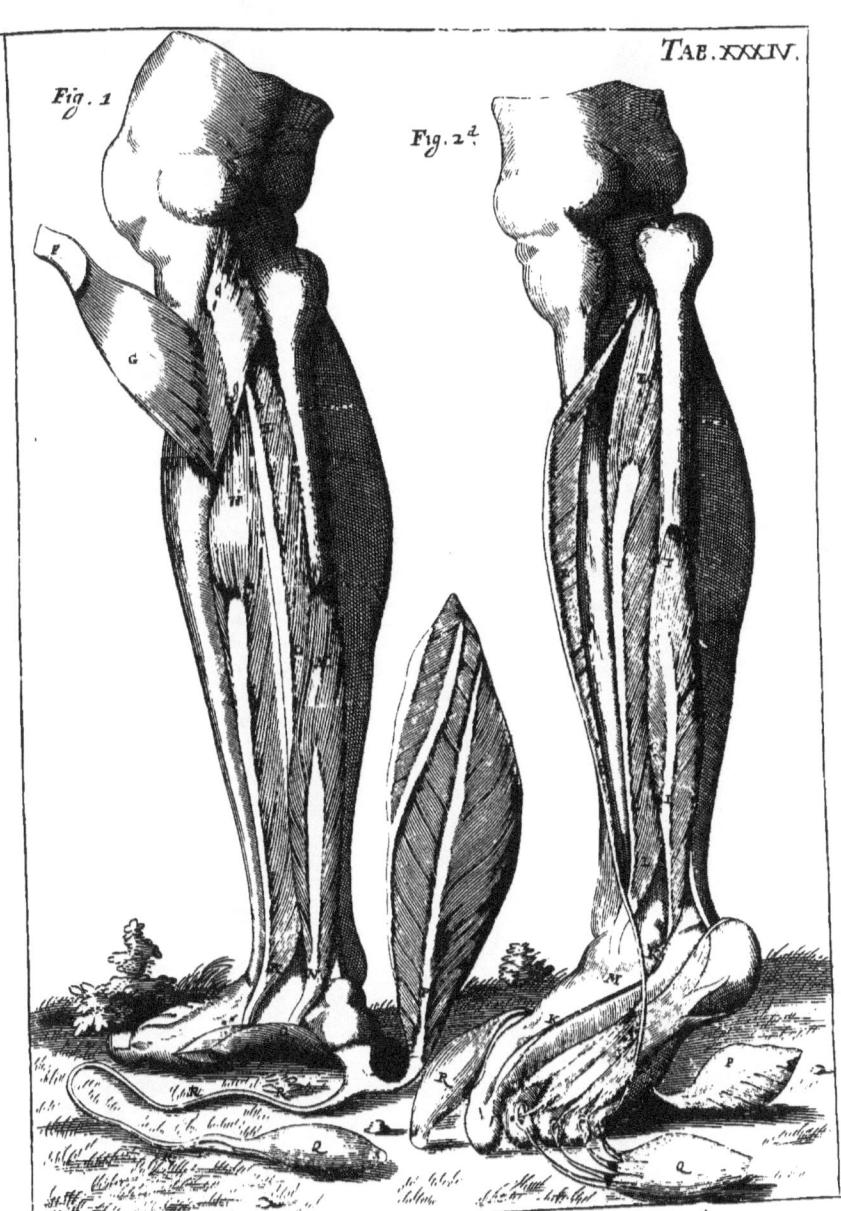

Flexores Primi Internodii Digitorum, seu Lumbricales.

These are called *Lumbricales* both from their Use, Figure, and Origination, much resembling Earth-Worms being put into the Hand. *These bends the first Joynts of the lesser Toes.*

They do arise round and Fleshy from the Tendons of the *Perforans* and *Perforatus*, and are inserted by small Tendons laterally to the first Joynts of the Toes; if you examine this well, you will find them to receive an addition of Carnous Fibres from a Musculous Flesh seated or implanted in the inward Cavity of the *Os Calcis*, and proceeding Fleshy half the Sole, there doth make these; and from it sometimes doth proceed a Tendon to the first of the Toes, and makes *Secundi Internodii*; and you shall also find that the Tendon of the *Tertii Internodii Flexor* doth run through the Body of this, and is not perfectly divisible.

These you have at *Tab.* 37. *Fig.* 1. at *K. K. K. K. F. F. F. F.* Shews the same at *Fig.* 2. *ejusd. Tab. R. R. R.* Shews their Fleshy Mass laid bare at *Tab.* 36. *Fig.* 1. *S. S. S. S.* Shews the Tendons whence they do arise, *T. T. T. T.* Shews their four Tendons.

Adductor Pollicis Major.

This brings the Toes inwards.

THis ariseth Nervous from the inward part of the Heel, and from the Ligament, which keeps that and the *Talus* together, and so growing Fleshy and round, becomes Tendinous, inserting it self obliquely into the lateral and inward part of the first Joynt of the Great Toe.

Use. This abduceth the Great Toe from the rest of the Toes; or, as I may say, draws it inwards.

This you have at *Tab.* 37. *Fig.* 1. at *E. F. F.* Shewing its Tendon; At *Fig.* 2. *ejusd. Tab.* you have the same laid bare at *Q. Q. M.* Also shews the same at *Fig.* 1. *Tab.* 35. laid bare, *O.* Shews this at *Tab.* 34. *Fig.* 1. *R.* Shews the same laid bare, *Fig.* 2. *ejusd. Tab.*

Abductor. Minimi Digiti.

THe Little Toe hath a particular *Abductor* allowed it, *This abdu-* *ceth the little* from the Heel, arising Nervous from its External part, *Toe from the* and growing Fleshy in its progress, runs with a small *rest.* Fleshy Tendon under the *Os Metatarsi*, which is immediately plac'd before the Least Toe, and so is inserted into the External side of its first Joynt.

This you have at *Tab.* 37. *Fig.* 1. at *G. G. H. H.* Shews its Tendon, *I. I.* Shews the same laid bare at *Fig.* 2. *ejusd. Tab. L M.* Shewing its Tendon, *K. K.* Shews its beginning, *Q.* Shews the same at *Tab.* 35. *Fig.* 1. *P.* Shews this at *Tab.* 34. *Fig.* 2.

Addu

Adductor Minor, five Transversalis Placentini.

This draws the great Toe to the little Toe.

THis by *Cafferius*, who first found this out, gave it the name of *Transversalis*, because it binds the first Joynt proceeding from the Ligament of the Little Toe, and is carried transversly Fleshy, and marcheth out with a short and broad Tendon inwardly, to the first Bone of the Great Toe.

Use.

The Author of this Muscle doth assign this Use to it, That drawing the Great Toe towards the Little one, it makes a hollowness in the Foot; so as that in unequal and stony places, an apprehension as it were being hereby made, we may tread and walk more steadily, and commodiously, this being as a Ligament to the Foot, to keep it from slipping or sliding, and for a more secure ambulation; for by the help of this Muscle, the Foot is brought into that Figure, that it makes the step sure, and as it were doth apprehend the part it is set on.

This you have at *Tab.* 37. *Fig.* 4. at E. laid bare, H. Shews the same at the Letter *A.* the first Toe, P. Shewing the same at *Tab.* 36. *Fig.* 2. P. Shews the same at *Tab.* 35. *Fig.* 1.

Tibiæus

Tibiæus Anticus, sive Musculus Catenæ.

THis is the first of the Inflectors, whose Tendon being *The bends the Ankle.* transversly dissected or amputated, the Patient is compelled to carry his Foot in a Sling, by the benefit of which he is able in his passage both to lift up and inflect his Foot.

This Muscle ariseth sharp and Fleshy from the uppermost Appendix of both the *Focils*, forwards: as also from the Ligament which binds them together, then being dilated, is narrowed about the middle of the *Tibia*, growing into a strong and round Tendon, running obliquely over the *Tibia*, and under the Annular Ligament, and is implanted into the inside of that *Os Tarsi* that is before *Os Pollicis*.

This Muscle I conceive governeth the Foot in motion, that it *vji.* doth not squail to much outward.

This you have at *Tab.* 35. *Fig.* 2. at *C. C. a.* Shewing its beginning, *b.* Its Termination and Tendon, *M. N.* Shews this at *Tab.* 36. *Fig.* 1. *O.* Shews its Tendon.

The Explanation of the Five and thirtieth Table.

FIG. I.

E.
- E. Tibiæus Posticus.
- e. Shews its Origination.
- f. Its Ten on.
- G. G. Peronæus Secundus.
- E. Shews its Tendon.
- M. Pollicem Adducens *laid bare.*
- N. O. Two *Interosseal Muscles.*
- P. Transversalis Placentini.
- Q. Minimum Digitum Abducens.

FIG. II.
- C. C. Tibiæns Anticus.

- a. Shews its beginning.
- b. b. Its Tendon.
- H H. Peronæus Secundus.
- e. Shews its Origination.
- D. E. Its Tendon.
- I. I. Extensor Tertii Internodii Digitorum
- K. Shews its four Tendons.
- L. Extensor Secundi Internodii Digitorum.
- M. M. M. Shews its Tendon.
- N. Pollicis Extensor *laid bare.*
- O. Shews its Termination.

Fibulæ

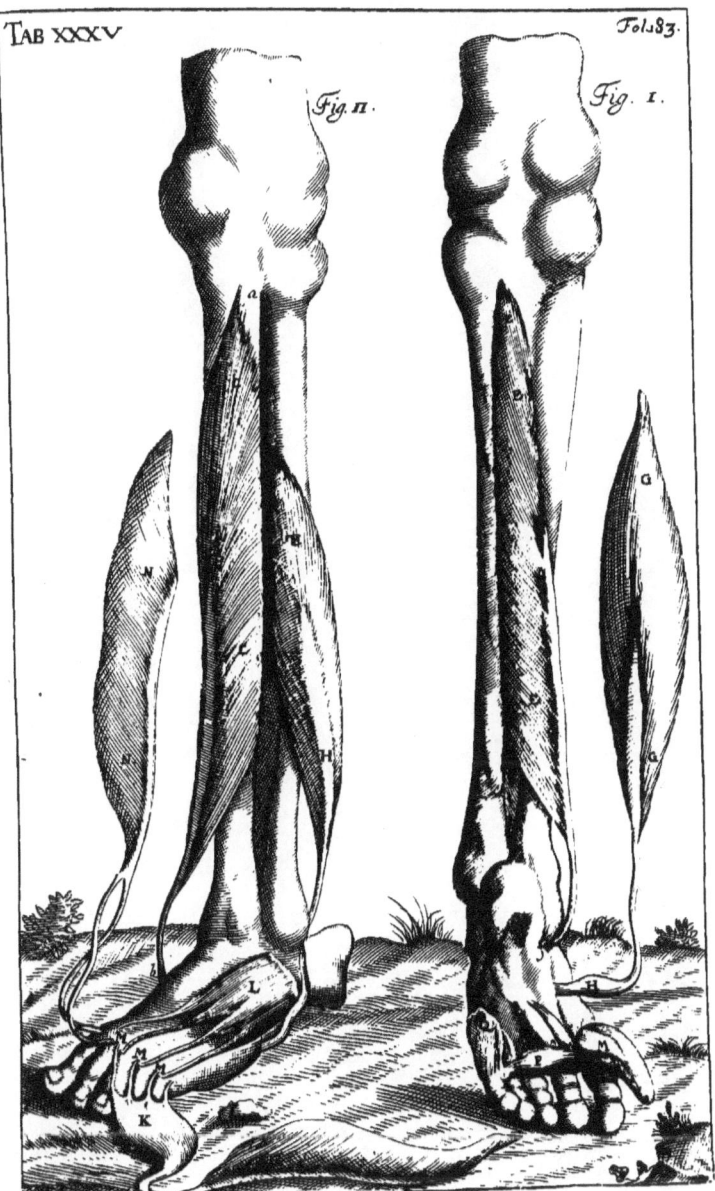

Fibulæus five Peronæus Primus, aut Posticus.

IT ariseth Nervous from the upper Appendix of the *Fibula*, and descending, it adheres to the outward part thereof, being externally round, inwardly livid, next the Muscle red, and marching half way, it becomes a Tendon running obliquely backward through the *Sinus* under the external *Malleolus*, and is inserted into the Root of the *Os Tarsi*, that adjoyns to the *Os Pollicis*. *This brings the Foot outwards.*

This Muscle draweth the Foot somewhat outward, and so regulateth it in Progressive Motion, that it cannot be cast too much inward. *Use.*

This you have at *Tab.* 36. *Fig.* 1. at *B. C. D. D.* Shewing its Fleshy Belly, *D*. Its Tendon, *V*. Shews the same laid bare at *Fig.* 2. *ejusd Tab. L. L.* Shews this at *Tab.* 34. *Fig.* 1. *L. L.* Shews the same at *Tab. ejusd. Fig.* 2.

Peronæ

Peronæus Secundus five, Semifibulæus, aut Anticus.

This bends the Ancle.

THis Muscle ariseth long and Fleshy from the *Fibula*, to which it adheres, and having made half its progress, becomes a round Tendon, running by the External *Malleolus*, and is implanted by two Tendons into *Os Metatarsi*, adjoyning to *Minimus Digitorum*.

This Muscle is also called *Semifibulæus*, it being a near neighbour to the *Fibula*; As also by *Spigelius* is named *Bicornis* from its double Insertion into the small Bone adjoyning to *Minimus Digitorum*.

This you have at *Tab.* 36. *Fig.* 1. at *E. E. F.* Shewing its Tendon, *I. I. I.* Shews the same at *Fig* 2 *ejusd. Tab. K.* Shewing its Tendon, *G. G.* Shews the same laid bare at *Tab.* 35. *Fig.* 1. *H.* Shews its Tendon and place of its Insertion, *H. H.* Shews the same at *Tab.* 35. *Fig.* 2.

Pollicis Tensor.

THis ariseth Fleshy from the *Fibula* (or as *Vesalius* offer- *This extends the great Toe.* eth) from the outward side of the *Tibia*, where it parts from the *Fibula*, as also from the Ligament that joyns them, to which it strongly adheres, and so becoming a Tendon, runneth over the lower part of the *Fibula*, and under the transverse Ligament, and is inserted into the two Joynts, in the upper part of the Great Toe, and doth extend them di- *Use,* rectly; sometimes this Tendon is seen divided into two: one of which is inserted into the last Joynt of the Great Toe, the other into the *Os Metatarsi* which lies just under it.

This you have at *L. Tab.* 36. *Fig.* 1. *b.* Shewing its Tendon, *N. N.* Shews this laid bare at *Fig.* 2. *Tab.* 35. *0.* Shews its Termination.

The Explanation of the Six and thirtieth Table.

FIG. I.

B C. D. Peroneus Primus.
B. Shews its strong Origination.
C. Its Venter.
D. D. Its Tendon.
E. E. Peroneus Secundus.
f. Shews its Tendon.
G. Extensor Tertii Internodii Digitorum.
H. H. H. H. Shews its four Tendon.
I. Its fifth Tendon.
L. Pollicis Tensor.
a. Shews its beginning.
b. Its Tendon.
M. N. O. Tibiæus Anticus.
M. Its Origination.
N Its Fleshy Venter.
O. Its Tendon.
S. Extensor Secundi Internodii Digitorum.

FIG. II.

D. D. D. Peroneus Primus.
E. Its Tendon.
I. I. I. Peronæus Secundus.
K. Shews its Tendon.
N. Pollicis Tensor *laid bare.*
P. Transversalis Placentini.
R. R. T. T. Lumbricales *laid bare.*
R. R. R. The Fleshy Mass from whence they do arise.
S S. S. S Shews the four Tendons whence they do arise.
T. T. T. T. Shews their own four Tendons.

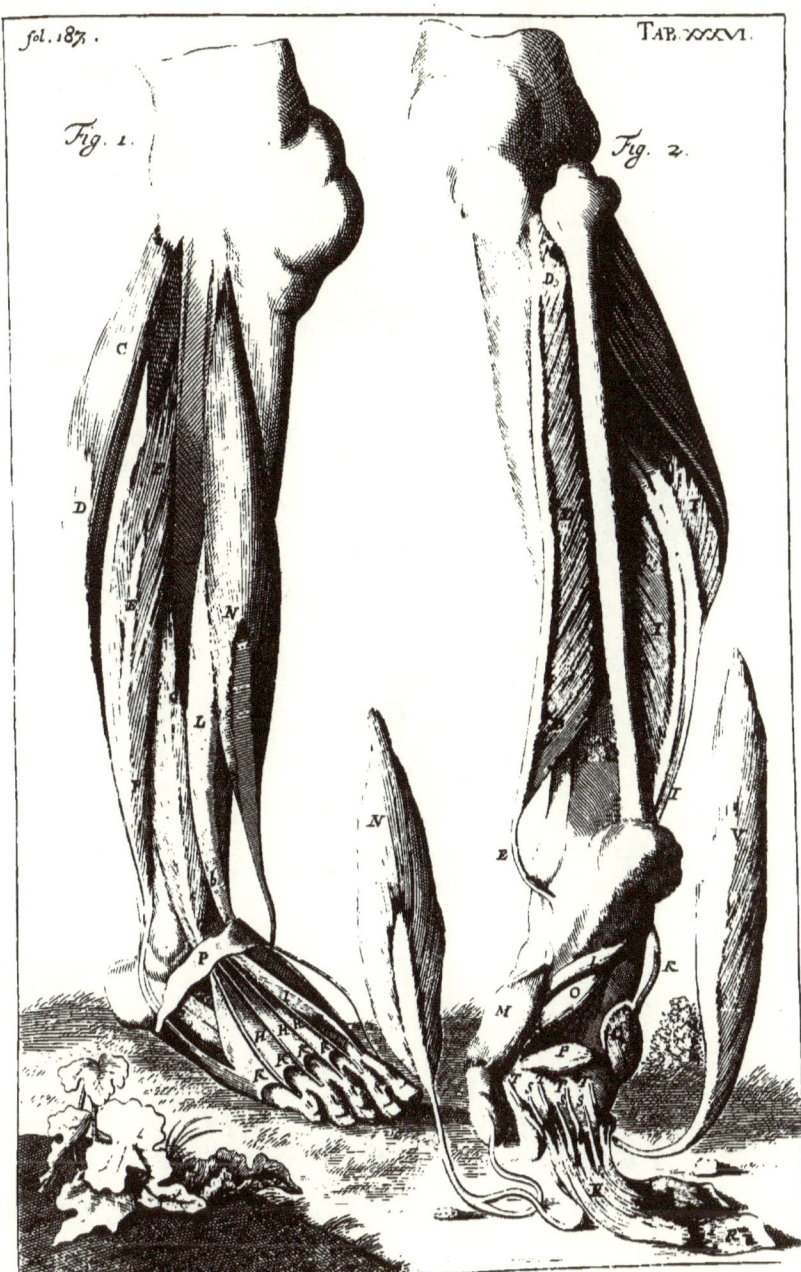

Extensor Tertii Internodii Digitorum, aut Longus.

This ariseth with a Nervous and Fleshy beginning, from *This extends* the outward, and forward Appendix of the *Tibia the third Joynts of* and presently becoming Fleshy, and adhering to the *the Toes.* Ligament that joyns the *Tibia* and *Fibula*, it descends directly according to the length of the *Fibula*, and passing under the Annular Ligament, it is divided into four Tendons, the *Use.* which do terminate in the upper part of the third or last Joynt of the four lesser Toes, and so extends them.

The other Tendons are tied amongst or between themselves by a certain Membranous Ligament, where they run through or over the back of the Foot.

This you have at *G. Tab.* 36. *Fig.* 1. *H. H. H. H.* Shews its four Tendons, *I. I.* Shews this at *Tab.* 35. at *Fig.* 2. *K.* Shews its four Tendons.

Exten-

Extensor Secundi Internodii Digitorum, aut Brevis.

This extends the second Joynt.

THis ariseth broad and Fleshy from the transverse Ligament upon the top of the Foot, and then appears, dividing it self into four several Muscles, which coming to the Toes, sendeth forth Tendons to the second Bone of the four lesser Toes, but chiefly to the second Joynts, where they intersect the Tendons of the former.

Use. These two are allowed to extend the four lesser Toes.

This you have at *Tab.* 35. *Fig.* 2. at *L. M. M. M.* Shews its Tendons, *S.* Shews the same at *Tab.* 36. *Fig.* 1.

The Explanation of the Seven and thirtieth Table.

FIG. I.

C.C. Flexor Secundi Internodii Digitorum.
D. D. D. Shews its four Tendons.
E. E. Pollicem Adducens in situ.
F. F. Its Tendon.
G. G. Minimum Digitum Abducens.
H. H. Shews its Tendon.
K. K. K. K. Lumbricales.
L. L. L. L. Their four Tendons.

FIG. II.

B. The Tendon of Flexor Tertii Internodi Digitorum.
C. Shews the Tendon of Flexor Pollicis.
D. D. D. Its Fleshy part.
E. E. E. Its four Tendons.
F. F. F. F. Lumbricales.
G. G. G. G. Their Tendons.
H. H. Two of the Interosseal Muscles.
I. I. Minimum Digitum Abducens.

K. K. Shews its beginning.
L. One Tendon of this Muscle.
M. M. Shews the other.
N. Flexor Secundii Internodii Digitorum laid bare.
O. Shews its sharp and Nervous beginning.
P. P. P. P. Shews its four Tendons.
Q. Q. Pollicem Adducens laid bare.
R. R. Two more of the Interosseal Muscles.

FIG. III.

C. C. C. C. Five of the Interosseal Muscles.
D. D. D. D. Other five of the Interosseal Muscles.
E. Transversalis Placentini.
F. F. F. F. F. Five of the Interosseal Muscles laid bare.
G. G. G. G. The other five laid bare.
H. Transversalis Placentini.

Interos-

TAB. XXXVII

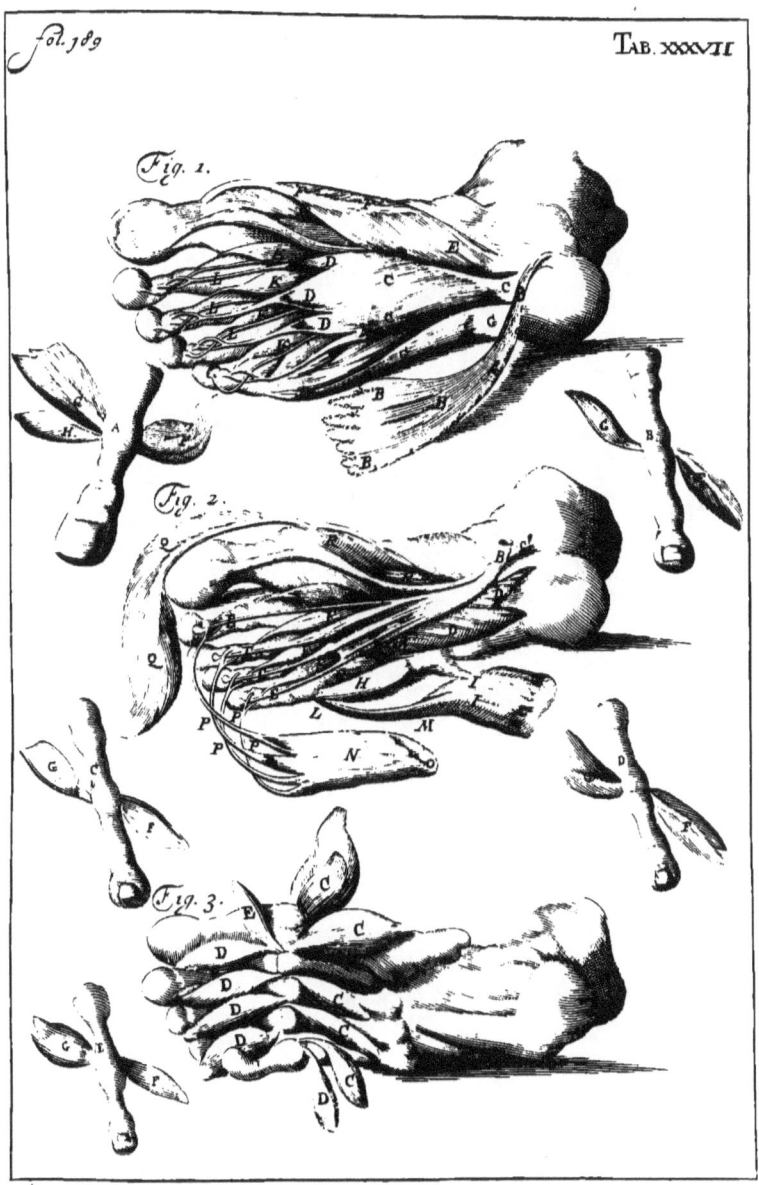

Interoffei.

THese have their names from their habitations, they *These move the Foot sl-* dwelling amongst the Bones; These Muscles do arise *liquely.* Fleshy from the sides of the Bones of the *Metatarsus*, and are inserted by some broad, but short Tendons, into the first Joynts of the Toes outwardly, but inwards to the second Joynts thereof. The outward working, they do abduce the Toes: the inward do adduce them; they both working toge- *Use.* ther, do extend them.

In the Sole of the Foot, which by some is called *Vestigium*, is a Fleshy Mass observable, the which like a Pillow or Bowl-ster doth keep warm the Tendons of the Muscles of the Foot, the which some Anatomists, although confusedly, do joyn with the *Musculus Transversalis*.

These Muscles you have delineated at *Tab.* 37. *Fig.* 3. at *C C. C. C. C. D. D. D. D. D.* And at *F. F. F. F. F. G. G. G. G. G.* in the Toes by themselves.

Thus have I concluded the whole Discourse of Muscles; all I shall farther add, is, the Enumeration of them, with their Re-ductions to their proper Places, Uses, and Offices, being either Relative to the Venters or to the Articulations, having already presented and drawn forth to you every of their Forms and Shapes: as also shewn their Originations and Terminations; leaving you Directions, how to find out every Muscle in its order by a proper Index made for that purpose.

F I N I S.

F f f   This

This Table sheweth the Reduction of the Muscles each to their proper place.

Use and Part.

THe Forehead is lifted up by *Frontalis*.

The Eyebrows,
- The upper { Lifted up by *Aperiens Rectus*.
  { Depressed by *Claudens Superior*.
- The nether lifted up by *Claudens Inferior*.

The Eyes,
- Rightly moved
  - Upwards by *Attollens*.
  - Downwards by *Deprimens*.
  - Inwards by *Adducens*.
  - Outwards by *Abducens*.
- Obliquely
  - Downwards externally by *Circumagens Exterior*.
  - Upwards internally by *Circumagens Interior*.

The Nose is
- Dilated by { *Primus Aperiens*.
  { *Secundus Aperiens*.
- Contracted by { *Primus Constringens*.
  { *Secundus Constringens*.

The Lips are
- Lifted up by *Attollens*.
- Drawn laterally by *Abducens*.
- Drawn down by *Deprimens*.
- Purs'd up by *Labium Constringens*.

The Cheeks are
- Drawn down by *Platysma Myodes*.
- Drawn inward by *Buccinator*.

The nether Mandible is drawn
- Upwards by *Temporalis*.
- Downwards by *Digastricus*.
- Laterally by *Masseter*.
- Forwardly by *Perygoides Externus*.
- Backwards by *Perygoides Internus*.

The Ears which are moved
- Externally
  - Upwards by *Attollens*.
  - Downwards by *Deprimens*.
  - Forwards by *Adducens*.
  - Backwards by *Abducens*.
- Internally { By *Externus*.
  { By *Internus*.

The Tongue is moved
- In Constriction, { By *Lingualis*.
  In Dilatation
- Forwards by *Geneioglossus*.
- Backwards by *Hypsiloglossus*.
- Upwards by *Myloglossus*.
- Downwards by *Ceratoglossus*.
- Laterally by *Styloglossus*.

The

(207)

The *Os Hyoides* is moved by
- Rightly
  - Upwards by *Mylohyoideus*, *Geneiohyoideus*.
  - Downwards by *Sternohyoideus*.
- Obliquely
  - Upwards by *Styloceratohyoideus*.
  - Downwards by *Ceracohyoideus*.

The Palate is
- Attolled by *Sphenopalatinus*.
- Depressed by *Pterygopalatinus*.

The *Fauces* are
- Dilated by
  - *Sphenopharyngæus Primus*.
  - *Sphenopharyngæus Secundus*.
- Contracted by
  - *Æsophagæus*.
  - *Cephalopharyngæus*.
  - *Stylopharyngæus*.

The *Larynx* is
- Dilated when the *Thyrois* is Extended by *Sternothyroideus*, *Cricoarytenioideus Anticus*.
- Contracted by *Hyothyroideus*.
- Shut
- Opened
  - While the *Arytenois* is
    - Contracted Directly by *Throarytenioideus*. Obliquely lateral by *Arytenioideus*.
    - Extended Rightly by *Cricoarytenoideus Posticus*. Obliquely laterally by *Cricoarytenoideus Lateralis*.

The Head is
- Contracted by
  - *Mastoideus* if both move.
  - Laterally if but one.
- Extended by
  - *Splenius* or *Triangularis*.
  - *Trigeminus*.
  - *Recti Majores*.
  - *Recti Minores*.
- Turn'd about by
  - *Obliqui Superiores*.
  - *Obliqui Inferiores*.

The Neck is
- Contracted by *Longus*, *Scalenus*.
- Extended by *Transversalis*, *Spinatus*.

The *Thorax* is moved
- Primarily by his proper Muscles which do
  - Dilate in breathing
    - Freely the *Diaphragma* alone contracted.
    - Coactively *Diaphragma* and *Intercostales Externi*.
  - Constringe in breathing
    - Freely *Diaphragma* alone relaxed.
    - Coactively *Diaphragma* and *Intercostales Interni*.
  - Extended as *Longissimus Dorsi*, *Semispinatus*, *Sacrolumbus*. which are retained in place by *Serratus Minor* (*Postici*), *Serratus Major*.
  - Contracted as *Musculi Recti*, *Obliqui Ascendentes*. *Abdominis*.
  - Turn'd about by *Transversi*.
- Secondarily by the Lumbal Muscles
  - Contracted by *Quadratus*.
  - Extended by *Sacer*.

The *Abdomen* is compressed
- Laterally by *Obliqui Ascendentes*, *Obliqui Descendentes*.
- Forwards by *Recti*.
- Downwards by *Pyramidales*, or Sometimes by *Transversi*.

The Loyns are
- Contracted by *Quadratus*.
- Extended by *Sacer*.

The

The Testicles are raised by *Cremasteres*.

The Bladder { Retains by *Sphincter Vesicæ*.
Excreates by { *Detrusor Urinæ*. *Pyramidalis*. *Obliqui Ascendentes Abdominis*.

The *Clitoris* is { Raised by *Musculi Graafiani*.
Depressed by *Musculus Labiorum uteri contractorum*.

The *Anus* is { Purs'd up by *Sphincter Ani*.
Elevated by *Levatores Ani*.

The *Penis* is impro- { Erected by *Erectores* or *Directores*.
perly said to be   Accelerated by *Accelatores*.

The *Scapula* is moved { Variously by *Cucullaris*.
Upwardly by *Levator Patientiæ*.
Backwards by *Rhomboides*.
Forwardly upwards by *Serratus Minor* } *Anticus*.
Forwardly downwards by *Serratus Major*

The *Os Humeri* is moved { Forwards by *Pectoralis*.
Upwards by { *Deltois*. *Octavus Humeri Placentini*.
Downwards by *Rotundus*.
Is carried about towards the { External part by { *Superscapularis Inferior*. *Superscapularis Superior*. *Nonus humeri Placentini*.
Internal part by *Subscapularis*.

The *Cubite* is { Extended by { *Gemellus Major*. *Gemellus Minor*.
Contracted by { *Biceps*. *Brachiæus*.

The *Radius* is { Pronated by { *Quadratus*. *Teres*.
Supinated by { *Longus*. *Brevis*.

The *Carpus* is { Contracted by { *Flexor Carpi Interior*. *Flexor Carpi Exterior*.
Extended by { *Extensor Carpi Exterior*. *Extensor Carpi Interior*.

The Fingers are { Contracted by { *Flexor Primi* *Flexor Secundi* *Flexor Tertii* } *Internodii*.
Extended by { *Primus* *Secundus* *Interossei* } *Extendentium Digitorum*.
Moved laterally by { *Interossei*. *Abductor Minimi*. *Abductor Indicis*.

The

(209)

The Thumb is
- Contracted
  - First by
    - Flexor Primi Internodii.
    - Flexor Secundi Internodii.
  - Secondly by
    - Primus
    - Secundus
    - Tertius
    - Quartus
    } Flexores Internodii.
  - Thirdly by *Tertii Internodii Flexor.*
- Extended by
  - Extensor Primus.
  - Extensor Secundus.
- Moved
  - Laterally internally by *Adducens.*
  - Outwardly by *Abducens.*

The Thigh is
- Extended obliquely
  - Backwards by *Glutæus Major.*
  - Forwards by *Glutæus Medius.*
- Contracted
  - Rightly by *Glutæus Minimus.*
  - Directly by
    - Psoas.
    - Iliacus Internus.
  - Obliquely by
    - Triceps.
    - Lividus.
- Moved about
  - Upwards by *Piriformis.*
  - Inwards by *Obturator Externus.*
  - Outwards by *Obturator Internus.*
  - Backwards by *Quadrigiminus.*

The Leg is
- Contracted by
  - Sartorius.
  - Gracilis.
  - Seminervosus.
  - Semimembranosus.
  - Biceps.
- Extended by
  - Membranosus.
  - Rectus.
  - Vastus
    - Externus.
    - Internus.
- Obliquely moved by *Suppopliteus.*

The Ancle is
- Extended by *Gasterocnemius*
  - Externus.
  - Internus.
- Contracted by
  - Tibiæus Anticus.
  - Peroneus Secundus.
- Moved obliquely lateral
  - Internally by *Tibiæus Posticus.*
  - Externally by *Peroneus Primus.*

The four lesser Toes are
- Contracted by
  - Perforans in the third
  - Lumbricales in the first } Joynt.
  - Perforatus in the second
- Extended by
  - Interossei in the first Joynt.
  - Secundi Internodii Tensor.
  - Tertii Internodii Tensor.
- Obliquely moved by
  - Interossei.
  - Minimi Digiti Abductor.

The Great Toe is
- Contracted by *Flexor.*
- Extended by *Tensor.*
- Obliquely moved by *Abductor.*

The first Joynts of the Toes are kept together by *Transversalis Placentini.*

The Skin
- Of the Sole of the Foot is moved by *Plantaris.*
- Of the Palm of the Hand by
  - Palmaris.
  - Caro Musculosa Quadrata.

G g g   An

An Alphabetical TABLE of the Names of the Muscles, with the Page cited in which each Muscle is treated on.

A.	Page		Page
Aperiens Palpebram Rectus	12	Claudens Nasum Externus	30
		Claudens Nasum Internus	31
Attollens Aurem	21	Coracohyoideus	42
Adducens aurem ad Anteriora	23	Ceratoglossus	51
Abducens aurem ad Posteriora	24	Cricoathyroideus Anticus	56
Abducens Nasi Alas	28	Cephalopharyngæus	59
Attollens Nasi Alas	29	Cricoarytenoideus Posticus	62
Abducens Labia	33	Cricoarytenoideus Lateralis	63
Arytenoideus	64	Cremastres	79
Anconæus	103	Cucullaris	88
Accelerator Penis	80	Cruralis	157
Adducens Pollicem, vel Adductor Pollicis Major	194	Caro Musculosa Quadrata	105
		Musculi Clitoridis	82
Abductor Minimi Digitorum Manus	121	Cervicalis Descendens	151
Abducens Minimum Digitorum Pedis	195	D.	
		Detrahens Aurem	22
Æsophagæus	57	Detrusor Urinæ	86
B.		Diaphragma	87
		Deltois	97
Buccinator	37	Digitorum Secundi & Tertii Internodii Tensor	128
Biventer	41		
Biceps Humeri	98	Dorsi Longissimus	149
Brachiæus Internus	100	Dorsi Latissimus	89
Biceps Femoris	176	Digastricus	41
Brachiæus Externus, vel Gemellus Major	101	E.	
C.		Extensor Cubiti brevis sive Gemellus Minor	102
Claudens sive Semicircularis Superior	13	Erector Penis	80
		Extensor Carpi Exterior	125
Claudens Semicircularis Inferior, seu Attollens	14	Extensor Carpi Interior	127
		Extensores Primi Internodii	135

Exten-

	Page		Page
Extensores Secundi Internodi Digitorum	204	**H.**	
Extensores Tertii Internodii Digitorum	203	*Hyothyroideus*	60
		Hypsiloglossus	53
Externus Tympani auris	25		

F.

I.

Frontalis	11	*Inferior sive Attollens Semicircularis*	14
Flexor Carpi Interior	106		
Flexor Carpi Exterior	107	*Internus Tympani Auris*	26
Flexor Secundi Internodii Perforatus	108	*Intercostales Exte. ni*	76
		Intercostales Interni	77
Flexor Tertii Internodii Perforans	109	*Interossei Manus*	124
		Indicem Abducens	133
Flexor Secundi Internodii Pollicis	110	*Iliacus Internus*	159
		Interossei Pedis	205
Flexor primus primi Internodii Pollicis	114	*Iliacus Externus*	163
Secundus	115	**L.**	
Flexor primus secundi Internodii Pollicis	116	*Labium Inferius Deprimens*	34
Flexores primi Internodii sive Lumbricales	117	*Par Labium constringens*	35
		Lingualis	55
Flexores Digitorum Tertii Internodii Perforantes	187	*Longus*	70
		Levatores Ani	83
Flexor Pollicis	190	*Latissimus sive Aniscalptor*	89
Flexor Secundi Internodii Perforatus	191	*Longissimus Dorsi*	149
		Levator Patientiæ	91
Flexor primi Internodii Digitorum, seu Lumbricales	193	*Lividus.*	181

G.

M.

		Masseter	38
Geniohyoideus	49	*Mastoideus*	40
Gneoglossus	52	*Mylohyo dens Riolani*	48
Gemellus Major	101	*Myloglossus*	50
Gemellus Minor	102	*Minimi Digiti Abductor*	121
Gluteus Major	160	*Minimi Digiti Tensor*	129
Gluteus Medius	161	*Membranosus*	169
Gluteus Minor	162		
Gracilis	171	**N.**	
Gasterocnemius Externus	183		
Gasterocnemius Internus	185	*Nonus Humeri Placentini Obliquus*	95

	Page		Page
O.		**Q.**	
Obliquus Descendens	1	*Quadratus*	154
Obliquus Ascendens	3	*Quadrigeminus*	166
Obliquus Primus Oculi, vel Obliquus Minor	19	**R.**	
Obliquus Secundus sive Trochæus vel Obliquus Majorcum Trochlea	20	*Rectus*	173
		Rectus Abdominis	5
		Rectus Oculi Primus sive Elevator	15
Octavus Humeri Placentini	99	*Rectus Secundus sive Depressor*	16
Obliqui Superiores	146		
Obliqui Inferiores	147	*Rectus Tertius sive Adducens*	17
Obturator Internus	165	*Rectus Quartus sive Abducens*	18
Obturator Externus	167		
P.		*Rhomboides*	90
Pyramidales	7	*Rotundus Major*	92
Par Labium constringens	35	*Recti Majores*	144
Platysma Myodes	36	*Recti Minores*	145
Pterygoides Externus	68		
Pterygoides Internus	69	**S.**	
Pectoralis	72	*Sternohyoides*	144
P. lmaris	104	*Sternothyroides*	45
Pronator Radii Teres	111	*Styloceratohyoides*	61
Pronator Quadratus	112	*Styloglossus*	54
Primi Internodii Pollicis Flexor Primus	114	*Stylopharyngæus*	58
		Shpenopharyngæus primus	66
Pollicis Tertii ossis Tensor	120	*Sphenopharyngæus secundus*	67
Primi Internodii Extensores	135	*Scalenus sive Triangularis*	71
Pollicis Abductor	122	*Subclavius*	73
Pollicis Adductor	123	*Serratus Major Anticus*	74
Pollicis Tensor	201	*Serratus Minor Anticus*	75
Psoas	155	*Sphincter Ani*	84
Psoas Parvus	157	*Sphincter Vesicæ*	85
Secundi & Tertii Pollicis Tensor	132	*Superscapularis Superior*	93
		Superscapularis Inferior	94
Pyriformis	163	*Subscapularis*	96
Plantaris	184	*Secundus Flexor primi Internodii Pollicis*	110
Peroneus primus	199		
Peroneus secundus	200	*Secundi Internodii pollicis Flexor*	
Pollicis Flexor	190	⎧ *Primus*	116
Pectoralis Internus sive Triangularis	78	⎨ *Secundus*	117
		⎩ *Tertius*	118
Pterygo-palatinus	61	*Quartus*	119
		Supinator	

(213)

	Page		Page
Supinator Radii Longus	139	Temporalis	39
Supinator Radii Brevis	134	Thyroarytenoides	65
Serratus Posticus Superior	137	Trigeminus sive Complexus	140
Secundi & Tertii Pollicis Tensor		Transversalis Cervicis	141
	132	Triceps	180
Serratus Posticus Inferior	138	Tibiæus Posticus	189
Splenius sive Triangularis	139	Transversalis Placentini	196
Spinatus Colli	142	Tibiæus Anticus	197
Sacrolumbus	150	Tensor Pollicis	201
Sacer	152	Triangularis	73
Semispinatus	153		
Sartorius	170	V.	
Semimembranosus	177		
Seminervosus	179	Vastus Externus	174
Subpopliteus	189	Vastus Internus	175
Spheno-palatinus	61		
T.		Z.	
Transversi Abdominis	9	Zygomaticus Riolani	32

F I N I S.

What Faults are committed by the Press, the Reader is desired to excuse, the Author's Occasions hindring him from so strict a Re-view of the whole Treatise from the Press.

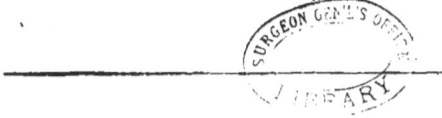

H h h